Neurodiversity

by John Marble, Khushboo Chabria,
and Ranga Jayaraman

Neurodiversity For Dummies®

Published by: **John Wiley & Sons, Inc.**, 111 River Street, Hoboken, NJ 07030-5774, www.wiley.com

Copyright © 2024 by John Wiley & Sons, Inc., Hoboken, New Jersey

Published simultaneously in Canada

For general information on our other products and services, please contact our Customer Care Department within the U.S. at 877-762-2974, outside the U.S. at 317-572-3993, or fax 317-572-4002. For technical support, please visit https://hub.wiley.com/community/support/dummies.

Wiley publishes in a variety of print and electronic formats and by print-on-demand. Some material included with standard print versions of this book may not be included in e-books or in print-on-demand. If this book refers to media such as a CD or DVD that is not included in the version you purchased, you may download this material at http://booksupport.wiley.com. For more information about Wiley products, visit www.wiley.com.

Library of Congress Control Number: 2024930702

ISBN 978-1-394-21617-8 (pbk); ISBN 978-1-394-21619-2 (ebk); ISBN 978-1-394-21618-5 (ebk)

SKY10074533_050624

Table of Contents

Introduction

You made it! If you're holding this book, or listening to it online, you probably have some big questions. Maybe you're wondering what being neurodivergent means, or if it applies to you or to someone you know. If you're a parent with a neurodivergent kid, you may be looking for tips on how to be the best parent you can be. And for those who are a spouse, friend, coworker, or anyone else close to someone who has a neurodivergent mind, you may be looking for ways to better understand their world.

Neurodivergent individuals often bring unique perspectives and problem-solving skills to the table. In fields such as technology, arts, and science, these diverse perspectives can drive innovation and lead to breakthroughs that may not happen in a more homogenous thinking environment. Embracing neurodiversity can be a significant asset in workplaces and communities.

The neurodiversity movement, however, emphasizes that regardless of the economic contribution potential, every human deserves to be seen, understood, and appreciated with respect and dignity as a unique creation of the universe. Learning about neurodiversity helps us challenge stereotypes and biases against people who think differently. It's a step toward creating a world where people aren't marginalized or misunderstood because their brains work differently. Understanding and accepting neurodiversity can lead to fairer, more equitable treatment for everyone.

It's completely normal to feel a mixture of curiosity, uncertainty, and even anxiety when it comes to understanding how the mind works. After all, no one gives you a handbook — but that's sort of why we wrote this book.

Neurodiversity For Dummies is your friendly guide to understanding the big, bold, beautiful world of neurodiversity. Throughout these pages, we explore what neurodiversity means, why it's important, and how it impacts you. We keep it simple, with clear explanations, useful tips, and real-life stories. And if you look closely, there's even a mention of Cher!

If you're wondering what Cher has to do with neurodiversity, well, then this book's for you. And she's not the only familiar face that shows up. You see, your life is filled with neurodivergent people, whether you realize it or not. Wouldn't it be amazing if you could understand and appreciate those people more? About 20 percent of us have brains that work differently in how we think, act, and experience the world. But until recently it was assumed that everyone's brains worked pretty much the same.

You're about to see how wonderfully complex our world truly is, and we explain it all in an easy and straightforward way. No need to stress over complicated words or theories; we've made everything simple and easy to understand. Feel free to read straight through or skip to the parts you find most interesting. So, find a cozy spot to sit, and let's get started!

About This Book

This book wasn't meant to sit on the shelf; it's a hands-on guide full of helpful tips, insights, and understanding. It's yours to really use. Go ahead and bend the pages, write on them, and highlight the parts that get you thinking. If you're listening to the audio version, feel free to just wave your hands, but be ready to write down the bits that strike a chord with you. Keep a pen close by — you may find a lot of things you want to remember. We explore key topics such as autism, dyslexia, ADHD, but we also demonstrate how neurodiversity extends way beyond these famous Big Three. There's lots of ins and outs of the human experience we get to explore.

If you're one of those parents or people who feel all alone, we acknowledge that pain. That's because we've felt that pain too. This book is a creation of neurodivergent people and family members working together. We've lived the expressions of joy, the moments of frustration, the feelings of triumph, and the tears of isolation that come from existing in a world that doesn't always get you.

We answer lots of questions, help you discover new things, and hopefully share some moments of laughter and self-reflection along the way. But this book isn't an information dump. It's not just a pile of facts. We mean it to be a conversation. No, we can't hear you. But we see you all the same.

Even if you picked up this book by mistake, let's still have that conversation. There's not a person on our planet whose life is not shaped and affected by neurodiversity in some way. Consider the phone you use, the songs you love, and the connections you have with family and friends. Every part of life is touched by our varied ways of thinking, shaded in subtle ways that aren't always immediately seen.

Neurodiversity is what happens when our colors blend together. It's a natural part of life; there's no changing that fact. We can't make more of it or wish it away. We only need to look around to see that it's there. Once we understand that neurodiversity is all around us, we need to choose how to respond. Should we be afraid of it, or welcome it with open arms? It may challenge us to think about how we act and treat others, including our friends, our children, and ourselves. But remember, this book is about moving ahead, not getting stuck in the past. It reminds us that we always have the chance to grow and change.

Foolish Assumptions

None of us are dummies. You're reading this book to get a clear, jargon-free understanding of neurodiversity. We don't talk down to you. We are especially cautious because many of us grew up being called names just for thinking differently.

Don't assume that your brain isn't normal simply because of the way that it thinks. And don't assume that other brains, which work in ways much different than yours, aren't just as normal. This book is your guide to understanding that all of us are equal. Our brains aren't problems to be fixed, nor are our different ways of thinking somehow "superpowers" to be praised. They're differences, that's all. They simply are what they are.

We don't assume that being different doesn't bring challenges. Lots of neurodivergent people live life in ways that are tough. But they also laugh, giggle at silly things, lose their train of thought, and hope for future things. To put it bluntly, they're just like everyone else. And if you're not one of these folks, what are the things that cause you pain? Is it feeling left out, struggling with your own problems, or sometimes worrying about the future? See? We're all the same.

Some parts of this book are written for specific groups, such as parents, teachers, or neurodivergent individuals. However, the lessons in these chapters are useful for everyone. We all want to be recognized, understood, and valued for who we are. To achieve this, we need to show the same understanding and appreciation to others that we hope to receive ourselves.

We also recognize that the language we use to describe neurodiversity is always evolving and varies across cultures. So, we've chosen to use words that are clear and easy to understand for readers from many different backgrounds. We've specifically opted for neutral words like *condition* to avoid negative connotations and to foster a more inclusive and understanding approach to all humans. No matter who you are, or why you're here, we hope you'll take away a better understanding and appreciation of neurodiversity by engaging with this book.

Icons Used in This Book

Throughout the book, we have used a set of icons in the margins to highlight the most critical things we want you to take away.

REMEMBER

When you see this icon, you know that the information that follows is important enough to read twice! Information in these paragraphs is often conceptual.

TIP

This icon indicates practical information that often translates key concepts into actionable advice.

WARNING

This icon highlights information that may be detrimental to your understanding and actions if you ignore it. We don't use this one much, so pay attention when we do.

Beyond the Book

Keep in mind that neurodiversity is a wide-ranging topic. We cover as much as we can in this book without overwhelming you. But no one book can cover every detail about neurodiversity and individual neurodivergent conditions. But don't worry! In addition to the abundance of information and guidance we provide in this book, you get access to even more help and information online at Dummies. com. Just go to www.dummies.com and search for "Neurodiversity For Dummies Cheat Sheet."

Where to Go from Here

We recommend that you start by browsing the detailed table of contents and then going straight to the chapters you are drawn to. You don't need to read this book from start to finish. Think of it as a buffet — you're free to pick and choose what interests you most. If you're in a rush and need some quick takeaways, the final part of the book is ideal. It neatly sums up main ideas and practical tips. But, if the concept of neurodiversity is new to you, we suggest you start from the beginning.

As you become more and more familiar with neurodiversity and venture into practicing the suggestions and recommendations from the book, you may want to return to different sections of the book to re-read them. We promise you will discover new information and experience many eureka moments along the way. However you choose to read this book, we hope you'll use it to discover that life is much more varied and beautiful than you think. We thank you for letting us be a part of your journey toward neuroinclusion.

1

Understanding Neurodiversity

Understand that neurodiversity is normal and has always been part of human history.

Discover how neurodiversity and disability overlap and that disability is nothing to be feared.

Deepen your appreciation for the neurodivergent people around you — and if you're neurodivergent, understand that there are many others out there like you.

Examine the common barriers to forming a completely neuroinclusive world where every human thinking style is understood, appreciated, and accommodated, and how to help overcome them.

Chapter **1**

What Is Neurodiversity?

You probably have a lot of questions if you picked up this book and turned to this chapter — whether those questions are about yourself, your child, or someone you love. Maybe you're feeling overwhelmed, or confused, or perhaps you're struggling to learn how to empower yourself. We get it. Life is a complex, often bewildering, experience.

There is something big happening in the world, and you're right in the front row to see it. What we long thought we knew about humans — how we think, act, and form connections with others — turns out to be an incomplete picture at best. That perspective had a lot of good intent, but it also had faulty misconceptions, stereotypes, and misguided ideas. It also did a lot of damage to a lot of people along the way.

Thankfully, we're now entering a world that has a better understanding of the complexities of our human species. In that understanding, the concept of neurodiversity plays a big part. We don't know everything just yet, but we're learning new things every day. And here you are, right at the front of it all — witnessing, participating in, and interacting with these new breakthroughs in human understanding. These new discoveries come as humanity wakes up from our long-held assumptions, as we begin to figure out who we truly are as a species and what our intricate complexities really mean. We're so glad you're here.

In reading this book, we hope you begin to understand *yourself* more deeply — whether you are neurodivergent or not. Perhaps you turned to this page as a parent. Maybe you're a teacher, a health care professional, a service provider, or a spouse. Perhaps you're a neurodivergent person yourself.

Of course, you may not be any of these things. You may have flipped to the first page of this chapter because you've long thought that "there's just something about my spouse (or my friend) that neither of us quite get, but we want to figure it out." You could also be one of many (many) others who may be reading this chapter and thinking, "Look, I don't know exactly what's going on inside me. All my life, certain things have been confusing and I don't know why." If things haven't clicked for you in life, the pages that follow could be your key to understanding why. No matter why and how you found this page, this book is for you.

There's a lot to discuss as we talk about neurodiversity. As we have this conversation, we share our insights and expertise on this topic. But we also press a bit further than most books typically do. We share with you our own experiences of living as neurodivergent people and raising neurodivergent children ourselves. We are honest about the worries and fears we've lived through, the obstacles we faced, and the many things we now wish we did differently because we know that many of you live through them too.

We also talk about the joy that we've found in accepting neurodivergence. We share how that acceptance has empowered us. And we discuss how that empowerment has improved our lives dramatically, allowed us to thrive, and has pushed us to continuously grow.

But first, we're going to take it slow. This is a big topic to wade into. So, we begin with this most crucial point, the one insight above all that we hope you take away from reading this book: Neurodiversity is normal, and so are you.

The Normalcy of Neurodiversity

Pause for a moment. Think about any family you may know. It could be the one you grew up in, the one you have now, or the family who lives across the hall or the street. Each member of that family is quite different — in age, in family role, and also in personality, quirks, and traits.

This mishmash of people all living together each has different talents, different challenges, different tastes, different personalities, different hopes, different dreams, and incredibly unique ways of perceiving and experiencing the world. Yet, each of these very different people forms part of one unit. Despite all their differences, they connect through similarities. Within the family unit, they cooperate, communicate, nurture, and support one another (and yes, often fight and disagree) as they learn, discover, and grow.

Just as with any family who lives together under one roof, every person on this planet forms a crucial part of our human family as well. We've always been a cooperative species, meaning that we've always needed each other and our differences throughout time. And living alongside one another, and benefiting from our differences, we have learned, discovered, and grown.

It doesn't matter whether you're a painter or a mechanic, whether you are short or tall, or whether you make your home in the Arctic or in the Amazon; we're all members of our human family. Actually, wait a minute. Strike that. It *does* matter whether you're a painter or a mechanic, it *does* matter whether you are short or tall, it *does* matter whether you make your home in the Artic or in the Amazon or anywhere else. It matters because *you* matter. You're not just an equal part of the human family; you're essential. Your friends need you. Your family needs you. We need you. The rest of humanity needs you too.

Your differences aren't a bug, they're a feature. Having differences is a normal part of being human, even if those differences are sometimes experienced as challenges.

Variations in the human condition

We guarantee that you're not alone in reading this book. A bunch of other folks are flipping through these pages too. Each of these readers varies in height, skin tone, eye color — you name it. But unless you're just a terrible person (and we like to think that you're not), you don't look at variations in height or eye color and think, "Whoa, that's super weird." These are just differences, that's all, and everyone has them.

We don't see these differences as bizarre because we've come to accept these variations in the human body as quite normal. But, here's the thing: It wasn't always that way. There were times in our human history when various members of our species encountered unfamiliar eye color, or height, or hair texture (or freckles!), and it really weirded us out.

The reason that many of us in the past found red hair, or green eyes, or dominant left hands disturbing is because humans are hardwired to be wary, and cautious, and slightly suspicious of the unknown. Throughout much of our history, this served us well at times. This instinct kept us from picking up snakes, eating potentially poisonous berries, or running up and hugging a lion with his cute, fluffy face. In fact, this evolutionary survival tactic is so deep rooted that we still carry it with us today.

But, here we are in the modern world. This instinct that we relied on for so long now seems to get in the way much more than it helps. It's the root of much of our prejudice and our exercises in discrimination, bullying, stereotyping, and many other cruelties we as humans tend to regularly do.

So, why do we no longer fear eye color or the number of freckles a person has on their skin, yet so many of us humans still fear conditions such as autism, attention-deficit/hyperactivity disorder (ADHD), and dyslexia? What's the difference here? Again, it's familiarity. There's nothing inherently bizarre about autism, ADHD, or dyslexia. It's in not knowing or understanding them that we fear. Part 2 of this book looks at each of these neurotypes in detail.

In the classes we teach to our neurodivergent students, we like to share a quote from two-time Nobel Prize winner Marie Curie (who herself expressed numerous neurodivergent traits). And here it is:

> There is nothing in life to be feared, it is only to be understood.

That's a key part of what we mean when we say "neurodiversity is normal." All humans have variations in the way that they think, feel, and experience the world — this is neurodiversity. Neurodivergent conditions such as autism, ADHD, and dyslexia (and dyspraxia, and dyscalculia, and many others) have been part of our human family for a very long time. While evidence suggests that some (even many) societies in the past did not fear them, our modern society — built on modern fears and modern preconceptions — is just beginning to understand these variations now.

As a society, we're slowly shaking off our prejudices to understand neurodivergent conditions as a normal part of the human experience. With each new insight we learn, each new discovery we make, we fear these conditions less and less. More importantly, the more we understand about them, the more we appreciate their necessity, how they enrich our society, and how they help our human family thrive. None of these conditions are to be feared — they're only to be understood.

Variations in the human brain

Think about the human brain. Now, there's no reason for us to expect that these couple of pounds of gray matter that are inside our head is going to be exactly the same from one person to the other. Thank goodness it's not. If it was, the world would be an extraordinarily boring place. Can you imagine if every one of us thought the same, learned the same, and acted in the exact same way? How dull! So, let's be grateful for that marvelous variety in how our brains function; it's an important part of what makes the human experience so rich and vibrant.

It's true that variations in the human brain are a normal aspect of our biological diversity. However, a lack of understanding about these variations has led to a lot of fear and anxiety around them. This fear often leads to stigma, isolation, and other barriers that those with these variations (which studies suggest are at least 15 to 20 percent of us) must face.

When we begin to recognize, accept, and respect these variations in human thinking as a normal, we begin to align with the concept of neurodiversity. And the impact of that alignment can be huge! It's like unlocking a door to a more inclusive and understanding world. We start to see neurological differences not as something to be fixed, but to be supported. It allows us to accept those neurologically different than us, enables us to understand their needs, allows us to include them more deeply in our lives, and helps us assist them in securing the accommodations and support they need.

Our lives are filled with neurodivergent people, whether we recognize that or not. We can't see their neurological differences the way we can see eye color or the freckles someone may have. But what we can notice is how these brain variations manifest in people's behavior and cognitive traits, such as how they communicate or socialize, how they deal with numbers and letters, or even how they react to light and sound. When we start paying attention to these things, that's when we can truly appreciate the full range of human neurodiversity.

Don't assume you can determine someone's neurodivergent condition just by casually observing their behavior or thinking patterns. Much of what you perceive in others is shaped by your own biases and incorrect assumptions. Making such judgments can result in stigmatization, reduced empathy, and missed opportunities. These consequences can not only harm both individuals involved, but also weaken the overall relationship between them.

"People often say to me, 'You don't look autistic!'" said video producer and fashion writer Rian Phin. "But, that's because your perception of what you think autism looks like is wrong." Phin — who is Black, known for her sustainable fashion aesthetic, and who has been profiled in both *Interview* magazine and *Vogue* — pressed the point further: "Think about it. How many autistic people who look like me have you seen in the media?"

Rian Phin hits a critical point. People have these preset ideas about what various neurodivergent people look like or act like. This is often shaped by limited or stereotypical media portrayals. (For how media representation can more accurately reflect the experiences of neurodivergent people, see Chapter 24.) This kind of narrow viewpoint presents a huge challenge for neurodivergent people and their families as they have sought to secure the understanding and support they need.

GETTING TO KNOW SOME HELPFUL TERMS

Throughout this book, we use the following terms. Their definitions are pretty easy to pick up as you go, but we thought it may be helpful to provide them all in one place.

- **Neurodistinct:** An alternative way of describing people who are neurodivergent. *Neurodivergent* and *neurodistinct* can be used interchangeably; although, neurodistinct places a stronger emphasis on the neurodivergent person's unique strengths.

- **Neurodivergent:** Having a brain that functions differently from the majority of people.

- **Neurodiversity:** The range of differences in brain function and behavior among all humans.

- **Neurofamily:** A family in which one or more members are neurodivergent.

- **Neuroinclusive:** An approach or environment that actively includes and accommodates people with diverse brain functions.

- **Neurological:** Related to the nervous system, including the brain and spinal cord, and how it influences behavior and functioning.

- **Neurotype:** A classification for a group of people whose brain functioning and associated traits are similar to each other. Examples of neurotypes include ADHD, autism, and dyslexia.

- **Neurotypical:** Having a brain that functions in a way that is similar to the majority of people.

- **Spectrum condition:** A neurotype category that includes various brain traits under one label, each person having a unique mix. It's not a straight line from a little to a lot, but more like a galaxy of stars (traits), where each individual has some of those stars but not all. Examples include autism and ADHD. The following figure illustrates a linear condition versus a spectrum condition.

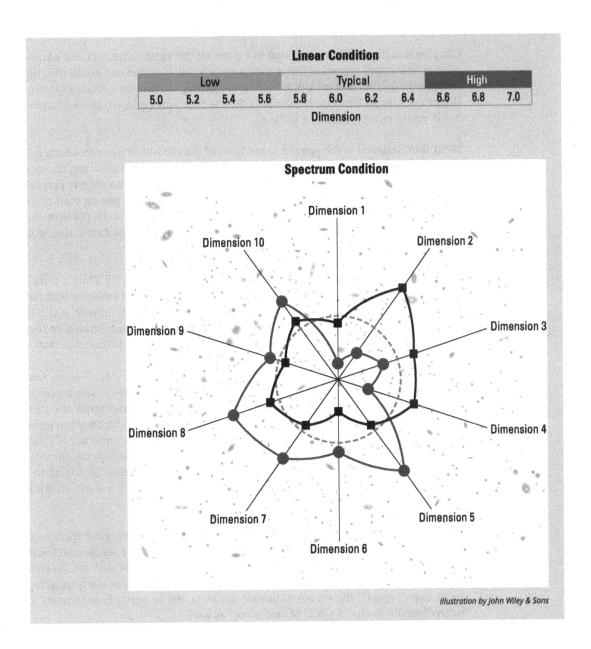

Linear Condition

	Low				Typical				High	
5.0	5.2	5.4	5.6	5.8	6.0	6.2	6.4	6.6	6.8	7.0

Dimension

Spectrum Condition

Dimension 1
Dimension 10
Dimension 2
Dimension 9
Dimension 3
Dimension 8
Dimension 4
Dimension 7
Dimension 5
Dimension 6

Illustration by John Wiley & Sons

How biodiversity informs neurodiversity

We've talked about how neurodiversity is an essential part of human diversity. Now, guess what? That diversity itself is a crucial part of the larger system of biodiversity that makes Earth the vibrant, living planet it is. While the term *biodiversity* itself was coined in the 1980s, the underlying concept has a much longer history rooted in the scientific understanding of biological diversity.

Our planet is *filled* with life, yet that life is not all the same. Scientists don't know exactly how many species of living things are around us. Estimates are all over the map, ranging from around 2 million to a staggering 100 million different types of life forms. Life on Earth is so incredibly diverse that pinning down an exact number is pretty much a guessing game.

What is understood is the general breakdown of the types living things which call our planet home. Most are insects (about 50 to 60 percent of all living things). Yep, we're a planet of bugs. Meanwhile, plants make up about another 17 percent of planetary species, and fungi — such as the mushrooms you put on your pizza or the yeast in your bread — clock in with a respectable 4 percent. Us noble mammals? Well, we only represent about 0.3 to 0.4 percent of life on Earth. Hey, size isn't everything.

REMEMBER

A diversity of lifeforms is necessary for life to exist. Earth may be a bug's world, but if *every* living thing on our planet were an insect, life would collapse. Without this diversity, we could be sweltering under thick blankets of sulfuric acid like Venus or we may find ourselves a cold and rocky place like Mars. Thanks to life, and it's mindboggling diversity, we can proudly say that we're better than that!

The rich variety of life that makes Earth a thriving planet is called *biodiversity*. And it is desperately needed. Why is it so essential? Well, a diverse set of species helps ecosystems be resilient. If one species goes down, others fill that niche and pick up the slack. Biodiversity also makes ecosystems more productive; they can grow more food, purify more water, and store more carbon. But that's not all! It's like an insurance policy against disasters and diseases. The more biodiversity there is, the better life on this planet can adapt and recover from challenges. So, maintaining biodiversity isn't just about saving the whales or the bees; it's about creating a stable, resilient planet for everyone.

Now, just because you're a mammal and in the minority of biological species on our planet, don't feel too glum. Our Earth needs you and your fellow mammals just as much as it needs the 50 to 60 percent of planetary life that are insects. You're just as normal as any bug. In fact, we know that you have many amazing traits that "typical" life on our planet do not have. We're not only supporters of neurodiversity, we're big fans of *biodiversity* as well.

The benefits of embracing neurodiversity

Just as biodiversity contributes to a healthy, resilient ecosystem, neurodiversity adds value and adaptability to human societies. Here are some key benefits:

>> **Problem-solving:** Different ways of thinking mean that problems can be approached from multiple angles. This increases the chances of finding innovative solutions.

>> **Creativity:** Neurodiverse individuals often have unique perspectives that can fuel creativity and artistic expression, enriching culture and even driving technological advancements.

>> **Specialized skills:** Some neurodivergent individuals may excel in specific tasks that others find challenging. For instance, an autistic person may be exceptionally good at pattern recognition, which could be invaluable in fields such as data analysis or programming.

>> **Resilience:** A neurodiverse community is better equipped to adapt to changing circumstances. Just as an ecosystem is more resilient when it's biodiverse, a human society is more robust when it includes a range of cognitive abilities and styles of thinking.

>> **Social development:** Neurodiversity can drive social change by challenging conventional ways of thinking and promoting empathy and understanding. This can lead to more inclusive communities that better serve the needs of all individuals.

>> **Efficiency:** When people are allowed to leverage their natural talents, regardless of neurotype, it often leads to greater productivity and efficiency in various kinds of tasks and projects.

>> **Quality of life:** Understanding and appreciating neurodiversity can lead to better mental health services, educational strategies, and career paths for neurodivergent individuals, improving their quality of life and ability to contribute to society.

REMEMBER

Neurodiversity includes everyone, not just people with neurodivergent conditions. If you're neurotypical, meaning your brain functions in a way that's common for most people, you're still a crucial part of human neurodiversity. Your way of thinking and experiencing the world is just as important as the perspectives and participation of neurodivergent individuals. So, don't think this topic doesn't concern you; we all need each other and we're all on this wild, wonderful ride through life together.

REMEMBER

Just as biodiversity is crucial for the health and stability of our planet, neurodiversity is beneficial for the well-being and progress of human societies.

Understanding Neurodivergent Conditions

Give or take, about 67 million people live in France. All are quite different from each other, yet all are quite alike. Each individual has their own strengths, their own personality, their own way of moving about the world. The life of a fashion designer living in Paris may appear quite different than that of a nun living in Marseille or a farmer outside Bordeaux. At first glance, you'd think, "Wow, they couldn't be more different!" But when you get down to it, they all share something in common: They're quintessentially *French*.

REMEMBER

A neurodivergent condition is kind of like being French. It's a label that groups together people who share many similar traits. But just as no two French people are identical, neither are no two people who share a neurodivergent condition. Each expresses traits associated with their condition differently and experience the world in their own unique way. That's totally normal. We're all individuals after all.

What is a neurodivergent condition?

To put it simply, a *neurodivergent condition* is a common variation in the human brain that's present in a large number of people. Those who share a neurodivergent condition have a different way of processing information, emotion, and stimuli compared to the majority of people (those with this typical brain type are often referred to as *neurotypical*).

Neurodivergent conditions include autism, ADHD, dyslexia, and dyscalculia that are present from birth and are part of a person for the rest of their life. These lifelong variations in the brain are sometimes referred to as *neurotypes*.

Think of it this way: You know how computers have different operating systems — Windows or Mac? Well, neurodivergent conditions can be thought of as different forms of human operating systems. Some naturally run Windows while others run Mac. All humans process information, generate thoughts and feelings, and interact with sensory input and social situations. We all do those things, but our neurological operating systems may do them in different ways — and that's okay!

REMEMBER

It's not a bad thing to be neurodivergent; it's just a different way of experiencing the world. And often, that difference comes with special skills and viewpoints that enrich our society in countless ways (for more on that, see Chapter 18).

Being neurodivergent isn't about being broken or needing to be fixed. It's a normal part of human diversity. Just as biodiversity makes ecosystems resilient and adaptable, neurodiversity does the same for human societies.

An older way of looking at neurodivergent conditions

You may be familiar with older terms for particular neurodivergent conditions. These include autism spectrum disorder, attention-deficit/hyperactivity disorder, dyslexia, dyscalculia, dyspraxia, and so on. If you are, you may have noticed something about these terms which stick out. Take another look: autism spectrum *disorder*, attention-*deficit*/hyperactivity *disorder*, *dys*lexia, *dys*calculia, *dys*praxia, and so on. Notice the *disorder*, *deficit*, and *dys* (originally from the Greek language, *dys* refers to something "difficult" or "bad") in these terms. Each of these terms suggests that something is broken, lacking, or defective.

REMEMBER

It's important to understand that these terms are not meant to stigmatize people. They originated in the medical field to describe specific challenges that neurodivergent individuals encounter, especially in a world that isn't always accommodating of different brain operating systems. However, these terms are increasingly considered outdated and limited, as they don't fully capture the entire experience associated with a neurodivergent condition.

WARNING

Now, we're not for a moment suggesting that the medical view of neurodivergent conditions is not valid. Nor are we suggesting that people with such conditions don't face difficulty or disability (for a deeper understanding of the intersection of neurodiversity and disability, see Chapter 2).

As with anyone else, neurodivergent people may face a wide range of health challenges, whether they are physical, psychological, or emotional. While some challenges may be linked to their neurodivergent condition, others arise from the stress of a person living in a world that is not readily accommodating of their needs. In addition, many neurodivergent people experience trauma because of misunderstandings, stigmas, and negative behaviors such as isolation, bullying, and abuse from those around them.

On our author team, John recalls a meeting of neurodivergent adults he once attended where the topic focused on addressing trauma. There, a retired military veteran asked that gathered group, "Why is it that the Veterans Administration provides me care for the trauma I experienced in combat, but no one ever talks about — let alone provides care for — the micro trauma that autistic people like me must experience and then deal with day, after day, after day as we go through life?"

So, when we say that traditional ways of describing neurodivergent conditions "are increasingly considered outdated and limited," we're not denying the particular medical challenges that neurodivergent people must face. What we're saying is that there is *much more* to a neurodivergent condition than the challenges a neurodivergent person faces by living in someone else's world.

Switching out our lenses to form a clearer perspective

As you go through life, how you view the world around you is called your *perspective*. It's extremely important. How you look at things informs your values, which then influences the actions you take. So, not only does your perspective affect your own experience, but your actions based on that perspective also influence how others experience the world.

On our author team, Ranga loves to share how humans often view the world around them through a pair of lenses. The first lens is *expectation* and the second one is *judgment*. Expectation refers to the preconceived notions or beliefs we have about how someone should behave, speak, or react in a given situation. Judgment involves forming an opinion or evaluation about someone based on their actions or words. Viewing people through expectation and judgment can cause unfair assumptions and missed opportunities for real connection. Often, we don't even realize we're doing it!

A neurodiversity mindset requires us to switch out those lenses. Instead of seeing people through the lenses of expectation and judgement, we begin to see people instead through the lenses of *inclusion* for abilities and *acceptance* for differences.

Including someone for their abilities means recognizing and valuing the unique skills, talents, and experiences a person brings to the table. It's about appreciating what someone can do, rather than focusing on what they can't. Part 5 explores how people with neurodivergent traits can thrive in all types of settings, including in education and the workplace, as well as in relationships and communities.

Accepting someone for their differences means embracing the unique qualities that make a person who they are, even if those qualities don't conform to societal norms or expectations. It's about appreciating the full range of human diversity, rather than trying to fit everyone into a one-size-fits-all mold.

Brain injuries, aging, and mental health

While the term *neurodivergence* is often associated with innate neurological differences such as autism or ADHD, it can also encompass conditions that arise later in life, such as traumatic brain injuries or the neurological changes that come with aging. This broadens the scope of neurodivergence to include anyone who experiences life with a brain that functions differently than what is considered typical, regardless of the origin of those differences. By including conditions such as brain injuries and age-related cognitive changes, the neurodiversity movement emphasizes the inherent value of all neurological experiences, not just those present from birth.

The term neurodivergence can also be applied to conditions that are not fully understood in terms of their origins or how they manifest, such as bipolar disorder, Tourette's, and obsessive-compulsive disorder (OCD). Even though the root causes and full range of symptoms for these conditions are still subjects of ongoing research, they are part of the human neurological experience. By incorporating these conditions under the umbrella of neurodivergence, we acknowledge that there is a range of "normal" when it comes to neurological function, and that understanding and acceptance should extend to all forms of neurodivergence.

Exploring the Neurodiversity Movement

The *neurodiversity movement* refers to anyone who values the acceptance, inclusion, and accommodation of neurodivergent people in daily life. There's no global headquarters, no political party, no organized ranks. It's simply made up of people who believe that all people are an equal part of the human family.

A concept emerges

The neurodiversity movement has its roots in the late 20th century, but it really gained momentum in the early 2000s. Originally, it was closely associated with the autism community. Judy Singer, an Australian sociologist, is credited with coining the term *neurodiversity* in the late 1990s. In 1999, Kassiane Asasumasu coined the term *neurodivergent* to describe individuals whose neurological functioning is different from neurotypical people. Singer's and Asasumasu's work set the stage for a broader understanding and acceptance of neurological differences. And it continued to be shared, shaped, and formed by all sorts of people who recognized that our world should be accepting, inclusive, and accommodating of people regardless of their neurotype.

A growing community

The Internet played a significant role in the growth of the neurodiversity movement, providing a platform for neurodivergent individuals to connect, share experiences, and advocate for themselves. In 1999, Asasumasu coined a second term, *neurodivergence*, to include neurological development and functioning that are atypical, diverging from the societal standards of "normal." Neurodivergence included far more than autism, and conditions like ADHD, dyslexia, and Tourette's, among others were added to the conversation. Online forums and social media became spaces where people could discuss the challenges and joys of being neurodivergent, often outside the medicalized narrative that dominated mainstream conversations. The movement began to influence academic discourse, public policy, and even employment practices. Companies started to recognize the unique skills and perspectives that neurodivergent individuals bring to the workplace, leading to specialized recruitment programs.

In recent years, the movement has been pushing for systemic changes in education, health care, and employment to be more inclusive and accommodating of neurodivergent individuals. (See Chapter 2 for a discussion of the disability rights movement.) This includes debunking myths and stereotypes and challenging stigmatizing language and practices. It has also developed stronger connections between neurodivergent individuals and parents of neurodivergent children — pushing for policies and practices that enable neurodivergent families to thrive.

Despite the progress, challenges remain, including stigmatization, lack of adequate resources, and unequal access to opportunities. But the movement continues to grow, driven by a committed community that values the positive contributions of neurodivergent individuals.

The goal of the neurodiversity movement

The neurodiversity movement aims for better representation of neurodivergent people in all areas of life, including in the media, in education, and in the workforce. It also fights for antidiscrimination laws and practical accommodations. The goal isn't just policy changes but also shifting how society views neurodivergence — from disorders to part of human diversity. It's all about inclusion and valuing different kinds of minds.

To put it simply: The neurodiversity movement hopes to shape a world where neurodivergent people and their families are accepted, integrated, and supported in all aspects of daily life. It's treating neurodivergent people, and those who care for them, like the normal people that they are.

Chapter **2**

Disability Is Not a Dirty Word

I s neurodivergence a disability? Answering that question is not as straightforward as you may think. That's because there are widespread misconceptions about what disability actually means. These misconceptions often generate anxiety and fear.

Many people find the word *disability* to be scary or taboo. But we can't talk about neurodiversity without understanding its connection to disability. In this chapter, we show you that disability is nothing to be feared. We discuss the misconceptions and fears that surround this word, define what a disability actually is, and then show you how neurodiversity and disability overlap. We also explore the disability community and how it empowers both the neurodivergent individual and those who may be caring for someone who is neurodivergent.

First Things First: Why We Fear Disability

Here's a quirky human trait: When we encounter the unknown, we tend to freak out. This tendency served us quite well in our ancestral days when encountering the unknown often represented a predator, a poisonous plant, or a rival human group. Fast-forward to today and this survival instinct shows up in some pretty amusing ways.

Consider that heart-stopping moment when you catch a glimpse of a garden hose from the corner of your eye as your brain initially mistakes it for a snake. Or why the shadows inside ordinary children's bedrooms suddenly transform into monsters at night. It's the same reason why some people get anxious about flying, even if they've never set foot on a plane, and why others enjoy the thrill of a scary movie or the adrenaline-fueled eeriness of Halloween. We're hardwired to fear the unknown.

So, what does this have to do with disability? It turns out, a lot. Just as with other unfamiliar things, it's not uncommon to find an irrational fear of disability among nondisabled individuals. That unease, which we refer to as *disability anxiety*, stems from a lack of understanding of what disability means.

Disability anxiety can make folks act . . . well, kind of weird. Here are some common ways disability anxiety manifests:

>> **Overcompensation:** Nondisabled people may go to great lengths to offer help, even when it's not needed or asked for.

>> **Infantilization:** People often speak to disabled people in an oversimplified manner, sometimes with a singsong voice (like how some people speak to toddlers and puppies).

>> **Fear:** Many people fear becoming disabled, which can lead to treating disabled people with pity, talking about disability in hushed tones, or avoiding the subject altogether.

>> **Fear of offending:** People may have a heightened worry about saying or doing the wrong thing, leading to awkwardness or hesitation in interactions.

>> **Overemphasis on disability:** A person's disability may be focused on to the exclusion of their other qualities, reducing them to a single characteristic.

>> **Underemphasis on disability:** Conversely, a person's disability may be downplayed with such phrases as "I don't even see you as disabled!"

What Disability Actually Means

When most people think of the term *disability*, they're actually referring to two distinct yet related concepts: disability and impairment. It's important to understand the difference.

Understanding impairment

An *impairment* refers to a variation in the body or brain that limits a person from doing certain activities. An impairment can be the loss of a limb, reduced vision, or a developed mental health condition. Some people are born with impairments and others develop impairments later in life.

Impairments can be temporary, or they can be long term. A broken arm is an example of an impairment because it requires you to limit certain tasks, and a sprained ankle is an impairment because it limits your mobility. Mobility can also be impaired through conditions such as spina bifida or multiple sclerosis, or through a spinal cord injury. A body after giving birth is impaired because it needs time to heal. Anxiety can also be an impairment when it limits life activities. We all experience impairments at various points throughout our lives. It's a part of being human.

Understanding disability

A *disability* arises when a person with an impairment is prevented from fully participating in their community due to their impairment or due to how society is structured around them. The following example examines how someone may be *disabled by an impairment.*

Imagine two coworkers, Olivia and Gary. Olivia is blind and Gary is not. Olivia and Gary have been working for the past month on an important presentation that they are to give to their team. They've nailed the details, checked the tech, and know their stuff. They are prepared.

The night before their meeting, Gary emails Olivia to say that his doctor has told him that he has the flu. Gary's body aches, he's nauseated, he's exhausted, his head is pounding, and he's running a high fever. There's just no way Gary's condition will allow him to come into work. Olivia will have to do the presentation without him. "No problem!" Olivia replies. She's prepared and can cover for him. The presentation goes great, and Olivia credits Gary for his contributions.

In this situation are two people who each have impairments. Oliva's impairment relates to her vision, and she's had that impairment since birth. Gary's impairment relates to the illness he is experiencing this week. In this case, which one of these two coworkers was disabled — or *not able* to complete the presentation due to the impairment they were experiencing? It was Gary. (Get better, bud!)

We'd argue that individuals are more likely to be disabled by how society is structured around them than by their own impairment. This view is shared by many disabled people and advocacy groups. It's also expressed in the Preamble to the

United Nations Convention on the Rights of Persons with Disabilities, ratified by at least 182 countries around the world, where it says, "disability results from the interaction between persons with impairments and attitudinal and environmental barriers that hinders their full and effective participation in society on an equal basis with others."

TIP

Recognizing that people can be disabled by other people is a powerful concept. It turns the attention from fixing the person (who most likely can't be "fixed" to begin with) to fixing the things that prevent them from participating fully in society.

This next example examines how a person may be *disabled by others*. Think about a person with a physical impairment named Ashanti. Ashanti was born with a condition that impairs her from walking on two legs. Instead, she uses a wheelchair to get around and go about her day. Her chair not only enables her to move from place to place, but also its battery charges her mobile phone, it has lots of pockets for her stuff, and it has a nice big cup holder for her iced coffee she orders on her way to work nearly every day. Pretty great!

Now, Ashanti arranges a business meeting with a company her firm is considering taking on as a client. When Ashanti arrives at the offices of the company, she is greeted with only a flight of stairs to get inside. For a person who uses a wheelchair, Ashanti is unable to enter the building. Instead, she calls her assistant to find another location and time where Ashanti and the company can meet.

Ashanti is disabled in this situation — she is *not able* to do an activity that she had planned on doing. Was Ashanti disabled by her impairment, or perhaps by her wheelchair? In this case, no. Ashanti was ready for the meeting, she was there, she was ready to go. What disabled Ashanti in this case was the exclusive use of a flight of stairs as an entrance.

REMEMBER

Don't worry about perfectly distinguishing between *impairment* and *disability* in every conversation. People often use *disability* for both, and that's okay. What matters is your intent and respect (for more on that, see Chapter 6). However, it's always helpful to mirror the language a disabled person uses to describe themselves.

TIP

You may have noticed that we use the phrase *disabled people* in this chapter rather than *people with disabilities*. Surveys and disability groups suggest that disabled individuals prefer the first term. Ironically, nondisabled people often use the latter, assuming it's more respectful. The most important thing is to not worry about semantics, but instead focus on the intent of respecting an individual's preferred way of describing themself.

Disability is common

At least one billion people on our planet currently live with a long-term disability according to the World Health Organization. That's a lot of us! And given the conservative methods used to determine that number, that's certainly an undercount.

Disability is normal

Disability is an inherent part of the human condition. It has been with us throughout human history, and disabled people have been part of every society. Disability is a natural element of human diversity, and all of us may experience disability temporarily or permanently at some point in our lives. Recognizing disability as normal isn't about saying, "Oh, wow! Disability is the best thing in the world!" It's simply seeing it as a reality in life that should be treated as ordinary and normal as any trait a person may have.

TIP

Remember when we talked about how disability is not about "fixing" an individual, but about modifying societal environments and attitudes to ensure equal opportunities for all? Well, it may be helpful to ask yourself, "What actions and attitudes that I have disable other people?"

Asking yourself that question can be quite empowering, and you don't have to just limit its frame to people with physical impairments. Are there things that you do to disable your coworkers, your family and friends, or strangers you may interact with through the day? Perhaps it's a grumpy attitude, or being rude, or parking in a vehicle spot reserved for disabled drivers because you're "just going to run inside real quick."

REMEMBER

You can't restructure the entire world, but you can control your own actions and attitudes and treat everyone around you with dignity and respect.

Neurodivergence and Disability

Disability and neurodiversity are fundamentally intertwined. Neurodiversity embraces brain function and behavioral differences — such as attention-deficit/ hyperactivity disorder (ADHD), autism, and dyslexia — as natural human variation, not pathological conditions. But it acknowledges that these differences can cause impairments in a society designed for neurotypical individuals.

An individual can be both neurodivergent and disabled — the disability emerges not from neurodivergence, but from societal structures unaccommodating of cognitive variations. An autistic individual may face disabling barriers due to unaccounted sensory sensitivities or communication differences, despite autism being a normal neurological variation.

This understanding can also help parents and caregivers of neurodivergent children replace fear and uncertainty around a diagnosis with appreciation of their child's unique world view. Instead of mourning lost "normalcy," understanding that their child's neurodivergence may cause them to face disabling barriers in society allows families to focus on advocating for their child's rights, seeking appropriate supports, teaching self-advocacy skills, and celebrating individuality.

The Disability Rights Movement

Disabled people have faced discrimination, prejudice, and limited access to education, employment, and public spaces for a long time. The *disability rights movement* is the name used to describe the groups of people who have come together to change that. Like other civil rights movements, it aims to break down barriers, advocate for the equal rights of disabled people, and create a more inclusive society.

The disability rights movement has played a significant role in shaping the way we understand and support neurodiversity. It's taught neurodivergent people the power of seeing themselves as normal and for standing up for their needs and rights. The movement has also shown all of us that it's more effective to focus on getting rid of the barriers that stop people from living life fully, rather than trying to change who they are. And laws that the disability rights movement has helped to secure — such as the Americans with Disabilities Act, the Canadian Human Rights Act, the Equity Act (U.K.), and the Disability Discrimination Act (Australia) — allow neurodivergent people to receive fairer treatment in many aspects of life.

THE STORY OF JUDY HEUMANN AND THE 504 SIT-INS

The 504 Sit-Ins marked a pivotal point in the disability rights movement. In 1973, Section 504 of the Rehabilitation Act, which prohibits disability discrimination in U.S. federal programs, was passed. However, the Department of Health Education and Welfare (HEW) failed to issue the necessary enforcement regulations year after year. This led to the disability advocates demanding that these regulations be issued by April 4, 1977. When that deadline passed without any action by HEW, nationwide protests ensued.

The San Francisco protest, led by Judy Heumann and Kitty Cone, stood out — it endured for 26 days! Protestors, including disabled people, their allies, and parents of disabled children, occupied the city's HEW office, garnering widespread media attention and pressuring the HEW Secretary. On April 28, the 504 Regulations were finally enacted, marking a significant triumph for disability rights. This 1977 milestone is considered the origin of the modern era of the disability rights movement by many disability advocates around the world.

You're Not Alone: Building Community

Doesn't it feel great when others just "get" you? Those who understand you, likely due to shared life experiences, create a sense of community. This human need for community — a network of individuals with shared interests or values — is pivotal for establishing a sense of belonging, fostering learning, and enjoying life. Our communities may vary from faith groups, neighborhood circles, book clubs, online forums, to sports teams. They may be gatherings based on shared hobbies, political beliefs, or just casual hangouts with friends.

Imagine your passion —a sports team, music, art, travel, or video games — being something you can't share with others. That would be awful, right? That's why we need community. Community is equally important for disabled people, as the connection with others who share their experience provides them a platform for ideas, understanding, and the realization that they're not alone. Disability isn't just about individual conditions — it fosters communities and cultures that significantly impact lives.

How the disability community empowers individuals

The disability community demonstrates that disability is much more than the particular impairment that a person may have. It is also people coming together to connect, support each other, and develop shared understanding and culture. This sense of unity and shared culture empowers individuals, giving them a stronger voice and sense of belonging in a community who truly understands who they are.

For neurodivergent folks, neurodivergent communities provide a connecting point for information, a space to share tips and solutions to life's daily challenges, and an opportunity to find mentorship on how to live a beautifully neurodivergent life within a neurotypical world.

Disability and neurodivergent communities come in many forms. These may include publications (*ADDitude Magazine* or *Spectrum Women*), advocacy organizations (Autistic Self Advocacy Network, Dyspraxia Foundation USA, and the International Dyslexia Association), social groups, or online forums. Some are focused on neurodivergent parents (Autistic Parents UK), people of color (The Autistic People of Color Fund), and women (Autistic Women's Alliance and Kaleidoscope Society). Others connect neurodivergent colleagues or professional peers.

Various neurodivergent communities actively advocate for rights, inclusivity, and societal acceptance, combating discrimination, and promoting a more accessible, equitable world. Leveraging shared knowledge and resources, the community serves as a beacon for newly diagnosed individuals to understand themselves, providing advice and direction on navigating their journey. (For more on neurodivergent community connections, see the appendix.)

How the disability community supports parents and caregivers

The disability community, particularly the neurodivergent subset, offers parents and caregivers of neurodivergent children valuable support, resources, and empowerment. Here, families can find an understanding space to share concerns, victories, and challenges without judgment. This shared wisdom provides both practical advice — such as navigating school accommodations and health care — and emotional support that's invaluable in tough times.

Parents and caregivers don't have to go at it alone. The autistic community has provided a good model in this regard. Websites such as the Thinking Person's Guide to Autism provide platforms for autistic adults and parents to share understanding and support, while groups such as Learn from Autistics help parents of autistic kids better understand their children by connecting them to autistic adults.

Moreover, neurodivergent communities provide a space for advocacy and change, encouraging parents and caregivers to defend their children's rights, combat discrimination, and promote inclusive policies across society. This connection can transform fear and uncertainty into resilience, advocacy, and change.

Chapter **3**

Someone You Know Is Neurodivergent

When Yuh-Line Niou first entered local politics, she had no idea that in a few years, she would be making a positive impact on families living more than 2,500 miles away. In 2016, Niou ran for a seat in her state's legislature — winning 76 percent of the vote. That victory made her the first Asian American elected to represent her district, which got a lot of media attention. But what many reporters didn't realize at the time was another amazing feat. Yuh-Line Niou had just become the first openly autistic person elected to the New York State Assembly.

In 2023, we visited Santa Clara County's child development offices in San Jose, California, to talk with parents of neurodivergent kids. There, taped to the meeting room wall, county workers had posted a display of Yuh-Line Niou for every family to see. It was something that Niou had no idea about. "I'm actually in tears right now," she said on the social networking site X in 2023, expressing her amazement upon hearing the news.

We naturally seek to learn from people similar to us, such as our parents, siblings, friends, or those with the same language, culture, or faith. This learning helps us grow, feel part of a community, and understand our world and our role in it. Picture a grandchild learning a recipe from a grandparent. It's not just about cooking; it's about connecting with family history and feeling part of a bigger story.

The people who posted the display of Yuh-Line Niou had never met her, but they understood a key point we hope you'll wholly embrace: There is a profound need for neurodivergent individuals and their families to see representations of neurodiversity in everyday life.

Too many neurodivergent people grow up thinking they're all alone. In this chapter, you discover how neurodiversity has always been part of human history and how it surrounds us every day. As you read these pages, we aim to deepen your appreciation for the neurodivergent people around you — and if you're neurodivergent, to help you realize that there are many others out there like you.

Neurodiversity Has Always Been Here

Imagine a world where everyone is exactly the same. Sounds kind of dull, right? Everyone has identical thoughts, feelings, and tastes in music, movies, and food. There's no variety, no debate, no new ideas to exchange.

Now, picture what happens next. It may seem peaceful at first, but without diverse thinking, innovation stops and problems are harder to solve. It's like eating your favorite food every day; eventually, it's not as enjoyable. Though sameness may feel safe at first, it results in a world without progress. Luckily, our world is not like that.

Discovering a tale as old as time

Neurodiversity isn't a new idea; it's been part of humanity forever. People with unique brain functions have always been around, greatly benefiting the world. They've produced stunning art, invented languages, pushed science forward, and explored new wonders and worlds (of the first three people to walk on the Moon, two were neurodivergent!).

"Neurodiversity may be every bit as crucial for the human race as biodiversity is for life in general," said writer Harvey Blume in a 1998 article published in *The Atlantic*. "Who can say what form of wiring will prove best at any given moment?"

Renowned professor of animal science Temple Grandin, who is autistic herself, agrees. Grandin asserts in her book, *The Way I See It* (Future Horizons) that without the presence of neurodivergence within the human gene pool, our history would look quite different. "You would have a bunch of people standing around in a cave, chatting and socializing and not getting anything done."

That's not to say that neurodivergent people are better than others. Instead, it highlights how diverse ways of thinking are critical for humanity's success.

Though we can't go back in time and diagnose historical figures, we can understand that many displayed neurodivergent traits. Included among these are artists such as Michelangelo, Leonardo DaVinci, Frida Kahlo, Andy Warhol, and Salvador Dali; scientists such as Barbara McClintock, Isaac Newton, Albert Einstein, Alan Turing, and Marie Curie; writers such as Maya Angelou, Hans Christian Anderson, Virgina Woolf, Lewis Carrol, Jane Austen, Emily Dickenson, and Agatha Christie; musicians such as Mozart, Tony Bennett, and Tom Wiggins; and world-shapers such as Benjamin Banneker, Harriet Tubman, and Mahatma Gandhi.

Seeing that neurodiversity means all of us

It would be a mistake to think that the role that neurodiversity has played in human history is only a tale about those with neurodivergent traits; it's a story that includes everyone, even those whose brains we call "neurotypical."

"I definitely look at people differently. I like to deconstruct, to pull a character apart, to work out what makes them tick, and my view will not be the same as everyone else," said autistic actor Anthony Hopkins when discussing his neurodivergent traits in 2017 in *The Daily Mail.*

Did you catch what he said? "My view *will not be the same* as everyone else." The two-time Academy Award winner isn't saying he's a better actor because of his autism or that he's a worse actor for it. Instead, Hopkins is emphasizing the importance of understanding himself and accepting his traits. Both neurodivergent and neurotypical people all have their roles to play.

Neurodivergence Is Rather Common

At least one in five people are neurodivergent (and this number may increase the more we learn about the brain). That's enough people for us to confidently say that neurodivergence is a rather common human trait.

Look at it this way: About 0.125 percent of humans are doctors, 0.74 percent are teachers, 2 percent have red hair, 10 percent are left-handed, and a whopping 20 percent of us are neurodivergent. That means the percentage of neurodivergent humans living on our planet is greater than the percentage of humans who live in Africa (17.5 percent), Europe (9.4 percent), North America (7.4 percent), South America (5.4 percent), or Australia/Oceana (0.54 percent).

Here's a brain teaser for you. Some researchers think that up to 35 percent of folks may be neurodivergent, while others think the percentage is even higher. Still more argue there's no real "normal" brain at all. But we'll stick with the current 20 percent estimate just to keep things simple.

Breaking down the statistics

If you're curious about how many people in the world belong to certain neuro-types, here's a quick rundown based on current estimates:

A little over 2 percent of the global population is autistic. Attention-deficit/hyperactivity disorder (ADHD) is seen in about 5 to 7 percent of us, and it's estimated that around 5 to 10 percent of humans are dyslexic.

Dyscalculia is found in 3 to 6 percent of humans. Dyspraxia is seen in about 5 to 6 percent of folks. And the estimates for dysgraphia vary widely, with it regularly being cited as being experienced by 5 to 20 percent of people (you can learn more about these variations in human thinking in Chapter 8).

Then there are conditions that are sometimes included in statistics about neuro-divergence and at other times not. They include conditions such as bipolar disor-der (2 percent of folks), obsessive-compulsive disorder (2 percent), Tourette's (1 percent), and others (you can read more about all of these in Chapter 9).

Understanding why statistics vary

When it comes to neurodivergence, statistics around the world vary depending on what country you look at. What we've given you are widely accepted global num-bers. So, don't be surprised if what is reported in your country differs from what you read here.

REMEMBER

Neurodivergent conditions don't have passports, and they don't gather in larger numbers in one country as opposed to another. Countries report different statis-tics due to diagnostic practices, societal awareness, and the level of resources applied to data collection and support. Statistics can also be influenced by cultural perceptions, health care policies, and the level of training and awareness among professionals.

WARNING

What we know about neurodivergence is always changing. So, we caution you to take the numbers we present here with a pinch of salt. The stats we've given you are widely accepted, and the go-to numbers used today, but that may shift over time. Just keep that in mind!

Understanding why numbers increase

In recent years, the percentage of people diagnosed as neurodivergent has increased. But this doesn't mean there are more neurodivergent people now than before. It's like how better telescopes show us stars that were always in the sky, even if we couldn't see them before. Likewise, we're getting better at recognizing neurodivergent traits that have always been there. That's a pretty great thing!

Reducing barriers leads to more accuracy

Reducing the barriers to diagnosis has allowed more people, especially people of color, girls, and adults, to get accurately diagnosed with neurodivergent conditions. We've come to realize that many earlier tests weren't really designed with them in mind, nor did they capture everyone's experience. Now, we're getting better at recognizing neurodivergent traits across a broader spectrum of people.

Parents learn about themselves when kids are diagnosed

It's also increasingly common for adults to seek their own assessment after their child is diagnosed with a neurodivergent condition. This often happens as parents see similar traits in themselves and want to understand why. A child's diagnosis can be like a moment of clarity, leading a parent to seek their own assessment. "Having two children on the spectrum helped me to discover who I truly am," said Morénike Giwa Onaiwu, a social scientist and researcher who is autistic, in an episode of her podcast, *Uniquely Human*. "It has been both illuminating and freeing."

The Role of Neurodiversity Today

Neurodivergent people are part of our everyday lives. They're our friends, family, coworkers. They may be you. But for a long time, people with brains that work differently were unfairly seen as less capable. Thankfully, society is now starting to appreciate and value these differences more. We're not yet where we should be, but we're definitely moving in the right direction!

"In recent years, we've slowly begun to see increased awareness and acknowledgment that differences in how brains process and behave are, in fact, quite normal and common," said writer and acclaimed nutritionist Ariane Resnick in a 2021 article for Verywell Mind. She continues: "While we had previously thought of neurodivergent traits as 'illnesses' or 'diseases' . . . these conditions come with equally important benefits."

Neurodiversity is great for society. People with different brain types are often really good at many things the world desperately needs. They bring unique skills and ideas, making communities better. When we accept and celebrate these differences, it enables us to spot new insights, solve more problems, and create things that work better for us all.

Understanding its role in our culture

Neurodiversity plays a crucial role in our culture by shaping our collective identity. Much of what we think about ourselves as humans has been shaped by the perspective of neurodivergent minds. Neurodivergent individuals contribute huge chunks of talent and creativity to our art, music, literature, entertainment, innovations, and scientific practices.

While there's a common myth that neurodivergent people are best suited for areas such as science, technology, and engineering, the reality is that their talents are well-suited for just about any field. This includes teaching, manufacturing, agriculture, and the arts, to name just a few. All areas of human culture benefit from those who think differently.

"As an artist, I love flipping things around, turning them upside down to ask the 'what ifs?'," said Kai Syng Tan, an associate professor in arts and cultural leadership at the University of Southampton in a 2022 article in *Beshara Magazine*. "And as someone with ADHD, I love trespassing boundaries and borders, and colliding contrasting elements together to make new, novel connections." When it comes to art, it helps to think a bit differently. And here is where many neurodivergent people thrive.

Appreciating its role in the community

In our communities, neurodiversity fosters inclusivity and understanding. Different minds coming together create vibrant neighborhoods where people appreciate each other's strengths and unique contributions. By building connections across backgrounds, we strengthen the bonds that hold us together. It's in these interactions that we learn, grow, and support each other, making our communities more resilient and compassionate places to live.

Valuing its role in the economy

Neurodiversity is also a driving force in the economy. Diverse minds bring innovation and creativity to the workplace. Neurodivergent individuals often excel in problem-solving, attention to detail, and thinking outside the box. They contribute fresh ideas that lead to breakthroughs and advancements in various industries. When businesses recognize and support neurodiversity, they become more competitive and adaptable in our ever-changing world.

WARNING

Neurodivergent people should be valued for who they are, period. Some neurodivergent individuals may not be able to work, but they're still equal and important members of the community. While the contributions that neurodivergent people make to our economies should be applauded, our inherent worth as humans shouldn't be tied to how much we are able to work. Whether you have walked on the moon or require 24/7 support, we all have equal worth.

Extraordinary People: Ordinary Traits

Growing up in rural Sweden, Ingvar Kamprad often struggled inside the classroom. But his dyslexia made him a whiz at figuring out new things on his own — what we'd call "multisensory" and "hands-on" learning today. Thanks to his family encouraging these learning methods, Ingvar was ready to start his own business at the age of 17. Sitting down at the kitchen table, he filled out the necessary paperwork, and came up with a name. Using his initials, IK, he added the first letters of his family farm, Elmtaryd and his local village, Agunnaryd to create IKEA.

Ingvar Kamprad's dyslexia significantly shaped IKEA, including the flat-pack design of its furniture, and particularly its unique product names. Given his difficulty in remembering the numerical codes assigned to each product, Kamprad created an easy-to-remember system using the names of Swedish islands, people, and towns (a reason why you may be sitting on a *Karlstad* sofa or putting your feet up on a *Havsta* coffee table today).

It's not just Kamprad and IKEA. Plenty of neurodivergent people have achieved significant things. Yet, when their neurodivergence is discussed by others, it's often couched in a narrative of *overcoming* neurodivergence to find success. But is that an accurate way of looking at it? Not really. It turns out that neurodivergent traits are often the very thing that helps a person to excel.

"Dyslexic teachers and psychologists are common," states Rosa Kwok, a cognitive developmental psychologist at Birmingham City University in her 2020 TEDx talk, "Dyslexic? They're Not Broken." Kwok's research shows that dyslexic people often rise to become leaders in sectors like science and education, along with areas such as business, health care, and creative fields. And other research shows that autistic people contribute significantly to society's major innovations, while those with ADHD are twice as likely to become entrepreneurs than anyone else.

In pretty much any field you look at, you may notice that a lot of the top achievers have brains that work a bit differently than most. This includes business titans such as Charles Schwab, Michael Bury, Ted Turner, Feisal Nahaboo, and Barbara Corcoran; sports legends such as Ann Bancroft, Kevin Garnett, Michael Jordan, Greg Louganis, Terry Bradshaw, and Brittney Griner; actors and directors such as Jennifer Aniston, Daryl Hannah, Ryan Gosling, Keanu Reeves, Octavia Spencer, and Stephen Spielberg; and celebrated musicians such as Solange, Selena Gomez, Carrie Underwood, Dave Grohl, Stevie Wonder, Florence Welch, and . . . Cher. And that's just a few!

REMEMBER Neurodivergent traits aren't magical superpowers. They are ordinary characteristics, much like those of neurotypical people. And it's no secret that neurodivergent individuals often face challenges in traditional work environments (turn to Chapter 21 for more on that). But, when neurodivergent traits are accepted and understood, they often can contribute to a person's success.

Of course, even for the most successful among us, not understanding one's neurodivergence can present invisible barriers to personal and professional growth. In a 2023 article by Erika Janes for *Flow* magazine, actor Busy Philipps recalls the depression she experienced before being diagnosed with ADHD nearly two decades into her highly-praised career: "It's through the process of figuring things out and growing and experiencing and learning that you're able to come to the place where I feel like I'm at now, which is: I feel very confident and comfortable in who I am."

REMEMBER We don't expect you to memorize these celebrity names — don't worry, there's no quiz at the end of this book! We mention them because neurodivergent individuals often feel isolated, unaware of role models who share their own perspectives. Though most of us won't become famous, with support and acceptance, every neurodivergent person has the chance to do something amazing in their own unique way.

Chapter **4**

Why Is Our World Not Yet Neuroinclusive?

A little over a century ago we were living by lamplight, getting to work on foot or by horse. Now, we're in a world humming with technology. We fly to family gatherings, pull handheld devices from our pockets to watch movies, and globally connect to collaborate, create, and even explore the cosmos. Reflecting on our human journey, it's rather awe-inspiring.

Despite all of this progress, don't forget that there's still work to do. This is especially true when it comes to creating societies that are inclusive of neurodivergent people, who still face many barriers in the modern world. So, why isn't the world not yet fully neuroinclusive, a place where every human thinking style is understood, appreciated, and accommodated? In this chapter, we dissect and examine the current barriers to reaching that goal.

The Myth of Normality

The myth of normality assumes that there's one type of standard-issue human brain. This outdated myth is a major barrier to neuroinclusivity. If your way of thinking doesn't fit the so-called "normal" model, then society tends to treat you as odd.

In education

When it comes to education, the myth of normality not only takes center stage, but here it is the star of the show. Almost all modern educational systems are designed around this outdated myth. In Western societies, our contemporary educational frameworks are rooted in the 19th-century Prussian Model that mandated that every member of society attend school up to a certain level. This approach is tailored to meet the requirements of the Industrial Revolution rather than today's actual needs. Its emphasis lies on standardization, conformity, age-based grade levels, and rote memorization and regurgitation of facts. In such a system, students who think differently are often deemed "underachievers" or "troublesome."

The Prussian Model was adopted to instill conformity and maintain social control over Prussia's diverse citizenry. It promoted punctuality, obedience, endurance for prolonged tedious work, and a standardized curriculum. These skills were essential in an industrialized society and vital for preparing citizens to serve as potential soldiers, given Prussia's experiences in the Napoleonic Wars.

In many Western societies, the myth of normality has been so deeply integrated into our education systems that, for generations, students with different learning needs were segregated into special education classes. Despite good intentions, this led to stigmatization, isolation, reduced expectations, resource scarcity, and neglect of individual needs. While there's been a shift toward more inclusive practices, the myth of normality's influence still lingers here.

So, even in the 21st century our educational systems are stubbornly stuck in the past. Even though Prussia has not existed as an independent state since 1871, even though our society has evolved past the factory-based industrial era, and even though we don't need to prepare students anymore to defend us from possible invasions by Napoleon, these antiquated educational methods and the myth of normality still impede neurodivergent students and students from other marginalized communities today.

In therapy

In therapeutic settings, the myth of normality can turn therapy into a mission to "normalize" the client rather than work with the client's unique strengths and traits. It's all too common to see therapies focusing on making neurodivergent individuals fit into societal molds instead of empowering them to thrive as they are.

In both situations — education and therapy — the myth of normality creates an environment where neurodivergent individuals may feel misunderstood, excluded, or even deficient. And that's not good. Education and therapy should be about embracing and nurturing individual differences, not trying to replace them.

In the workplace

The workplace isn't immune to the myth of normality, either. Traditional work environments often favor a typical 9-to-5 structure, continual focus, and a certain style of social interaction. This can be hard for neurodivergent individuals who may operate best under different conditions. (We talk more about how our current workspaces and policies don't fully support neurodivergent employees in Chapter 21.)

Performance assessments and job roles often stress "normal" work behaviors, pressuring neurodivergent individuals to mask their unique ways of thinking and functioning. This not only generates stress but also deprives the organization of the diverse talents neurodivergent employees offer.

REMEMBER

Remember, the goal isn't to create an army of sameness. That's the realm of science fiction nightmares. Instead, as a society, we should focus on reality and create environments where every type of mind can thrive.

Why Things Are Designed to Look Like Us

We humans are pretty darn good at understanding and empathizing with experiences that look a lot like our own. This can be super useful in a lot of ways, but when we're put in charge of designing systems and societies, it can lead to a bit of a blunder.

This actually has a name, and it's a psychosocial phenomenon called the *mirror effect*. Picture this: When individuals set out to craft societal structures, they instinctively tailor them to their own needs, or those of people who navigate life in a similar way. It's not a deliberate snub to others; rather, it's the simple reality of how challenging it is to design for experiences you haven't personally lived.

As one example, consider city planning. In the United States, many cities are built around cars being the central transportation method. Makes total sense if you own a car, doesn't it? But this car-centric design unintentionally sidelines those who rely on public transit, biking, or their own two feet. No one set out to make life more difficult for those who don't own a car, yet that's the outcome.

The educational system is another example. School systems are usually structured around the belief that there is a universally correct way to learn. Those designing these systems often do so because they're replicating the educational systems in which they were raised. So, if your learning needs align with listening to lectures and taking tests, you're in luck. But what about students who thrive on hands-on activities? Or those who need to be physically active to process information? The system wasn't built with them in mind. (Once again, we blame Prussia.)

REMEMBER

Does this mean that those employing the mirror effect are evil, out to get those who aren't like them? Of course not. If we're being honest, we all do this a bit ourselves.

Suppose you're in charge of bringing dessert for a get-together, and you decide to bring your favorite triple-chocolate cake. But, to your surprise, not everyone is thrilled, as one friend dislikes chocolate and another avoids sugar. This too illustrates the mirror effect, where decisions are influenced by personal preferences, unintentionally overlooking others' tastes. It doesn't make you a bad person, but considering others' perspectives and offering different options can prevent this effect moving forward.

In the dessert example, your friends most likely were not bothered at all by your choice of dessert, but when the mirror effect is applied to large parts of our society, there can be unintentional and large-scale consequences. Without thinking and considering the experiences of others, a lot of people can get hurt.

REMEMBER

When crafting systems — from school meals to work structures — don't forget that "our way" isn't universal. By acknowledging this, we enhance the collective good.

Falling Into a Fixed Mindset

A "fixed mindset" is the belief that an individual's intelligence, abilities, and talent are unchangeable. Society routinely applies a fixed mindset when viewing neurodivergent people, falsely assuming that their brains aren't as malleable or adaptable and the neurotypical people around them — as if neurological differences inherently limit a person's capacity to learn, grow, or succeed compared to others.

Historically, this fixed mindset tore generations of neurodivergent people from their families, sending them off to live under lock-and-key in government-run institutions. Although there has been a shift in recent decades against institutionalization, this practice still continues in many places today.

But the restrictive effects of a fixed mindset don't stop at the gates of institutions. They permeate the everyday lives of all neurodivergent individuals, and the consequences can be severely limiting. For example, a person with ADHD may be labeled as perpetually unfocused and incapable of task completion, leading educators or employers to deny them opportunities for growth. An autistic individual, on the other hand, may be stigmatized as socially incompetent, leading to isolation and the belief that they're incapable of forming meaningful connections. (To see how a growth mindset can empower neurodivergent individuals, flip to Chapter 12.)

Acknowledging Barriers to Inclusion

We looked at how societal structures tend to be designed away from neurodivergent needs. We next delve into how such designs create real obstacles that prevent neurodivergent individuals from fully engaging in society.

Barriers society places in front of neurodivergent people

Following are examples of the barriers society routinely places in front of neurodivergent people:

>> **Education:** Traditional one-size-fits-all systems limit academic progress for neurodivergent students with diverse learning needs.

>> **Employment:** Workplaces lacking understanding and accommodations often lead to job insecurity, discrimination, and underemployment for neurodivergent individuals.

>> **Stigma:** Misconceptions about neurodivergent people can result in exclusion and bullying.

>> **Social expectations:** Neurodivergent people can often feel left out because they may not understand neurotypical social rules, or because neurotypical and neurodivergent people often don't fully understand their differences.

>> **Segregation:** Segregating neurodivergent individuals in education or work often results in social isolation and reduced community interaction.

>> **Institutionalization:** Historically, confining neurodivergent individuals to institutions has limited their freedom and social integration, exposing them to potential abuse.

- >> **Health care access:** Insufficiently trained medical professionals can limit effective access to health care for neurodivergent individuals.

- >> **Legal barriers:** Legal systems can fail to understand and accommodate neurodivergent needs, leading to unfair treatment.

- >> **Public spaces:** Public places often overlook sensory needs, limiting full participation.

- >> **Media representation:** The lack of accurate and diverse representation of neurodivergent individuals in media contributes to harmful stereotypes and misunderstanding.

Additional barriers neurodivergent people must navigate

A neurodivergent person must also navigate barriers deeply embedded within themselves. These take the form of beliefs often forced upon them in childhood and reinforced by others throughout their lives. Instead of living authentically, the neurodivergent individual often faces subtle or overt pressure to downplay their uniqueness. They're encouraged to conform, hide who they are, or replicate neurotypical behaviors. Other times, neurodivergent traits are ignored or simply dismissed. These are rather complex barriers that neurodivergent people often face.

REMEMBER

It's rare for the neurodivergent person to be taught who they truly are as they grow up. More often than not, neurodivergent children and their caregivers are not offered adequate information and support. They are left without a roadmap to comprehend and appreciate their differences, or a guide to understanding and advocating for their own needs. This is why it's crucial to support both neurodivergent children and their families, instilling an understanding and appreciation for their unique traits from a young age. Without this early education, many neurodivergent adults are left to navigate these complexities on their own later in life.

2

Recognizing the Types of Neurodivergence

Get a deeper understanding of the diverse world of brain variations, the reasoning behind doctors categorizing people with similar thought patterns, and when these groupings may not be accurate.

Explore how the autistic brain thinks and what it means to be autistic.

Find out about common ADHD traits and how to support and empower individuals with ADHD to make the most of its strengths.

Look at dyslexia, dyscalculia, dysgraphia, and dyspraxia, conditions that affect how people think, learn, and move, and that may make reading, writing, or math tough.

Explore additional neurological experiences that go beyond traditional categories of neurodivergent variations, including Tourette's, bipolar, OCD, cerebral palsy, and multiple sclerosis.

Look at neurodiversity through a framework of mental health.

Chapter **5**

Understanding Neurotypes

J ust as people have different tastes in music or food, people's brains also have their own unique styles of working. Some are neurotypical, which is like having a brain that follows the typical script society expects. Others have neurodivergent brains, which are like remixes of that script — they process, react, and interact in their own special ways.

In this chapter, you discover the diverse world of brain variations, including how neurodivergent individuals stand out from others and the unique strengths they bring. You also find out the reasoning behind why doctors categorize people with similar thought patterns, understand the benefits of these classifications, and recognize the times when these groupings may not be accurate. This journey will give you a deeper understanding of the neurodiversity spectrum and the rich tapestry it adds to human experience.

Introducing Neurotypes

You may have grown up thinking that all human brains are the same. After all, you can't poke inside the brains of others just to see whether their brain is similar to yours. However, scientists have discovered that about 75 percent of people do have a similar "type" of brain. So, you weren't that far off!

Scientists also estimate that about 25 percent of people have one of several naturally occurring variations of the human brain. These include conditions such as dyslexia, attention-deficit/hyperactivity disorder (ADHD), and autism. These variations are called *neurotypes*.

Your neurotype is your brain's unique style of processing information and the world around you — whether it's learning, thinking, or feeling. Neurotypical brains are like the standard version of a neurotype — they function in the way most people would anticipate. Neurodivergent brains, on the other hand, are like alternate versions. There's nothing wrong with them, but they have their own distinct way of operating. These neurotypes are often identified by doctors as autism, ADHD, dyslexia, among others.

WARNING

A word of caution! If someone has traits different than the majority of people, it doesn't mean that they aren't normal. Only 10 percent of people are left-handed, just 4 percent of humans live in the United States, fewer than 2 percent of us have red hair, and only 1 in 5 people living today owns a car. Each of these groups is *divergent* from the majority, but that doesn't mean they're not *normal*.

Understanding How Neurotypes Differ

You can't see variations in the human brain just by looking at someone. However, how we communicate, relate, and interact can reveal who we are. The most common areas of difference in neurodivergent individuals are in how we communicate, socialize, think, and explore the world through our senses. We discuss these differences in the following sections.

REMEMBER

At the end of the day, whether someone is neurodivergent or neurotypical, people are all pretty similar in the big picture.

Communicating thoughts

How you share information is called *communication*. The way you share information is unique to you, although you may communicate in ways that are similar to others. People within various neurotypes may communicate in ways that others may not understand. This may include unique body language, facial expressions, and speech patterns particular to one neurotype or another. People with dyscalculia or dyslexia may favor spoken language or visual communication over written numbers or words. Those with ADHD or autism may naturally communicate large amounts of information that others might struggle to process. The more we respect variations in communication, the more we can understand each other when we communicate.

Socializing with others

When you hang out and spend time with other people, that's called *socializing*. When people in one neurotype hang out with others like them, it can feel relaxed and easy. However, socializing outside of one's neurotype can often feel stressful or confusing. That's largely thanks to natural variations in our thinking and communication. We're still getting to know the ways we can make cross-neurotype socializing much easier.

TIP

Culture also shapes how we socialize. In some countries, it's easier for neurodivergent people to socialize with others because of local cultural norms. In other places, customs can make socializing more difficult for the neurodivergent person. For example, many autistic people find small talk (casual chitchat about mundane things) challenging and fake. Spending time with people in places where small talk is a crucial form of socializing can be tough. But in countries that place less emphasis on small talk, such as Finland and Sweden, it may feel easier for neurodivergent people to socialize.

Making sense of things

All the thinking, perceiving, and reasoning you do — conscious or not — is called *cognition*. It's how we make sense of things. Neurodivergent people use cognition in ways that are different from others. For example, they may

>> Think through large amounts of information quickly.

>> Rapidly examine information from multiple angles.

>> Think more deeply when processing information.

>> Interpret information and language more concretely.

>> Favor reasoning over intuition, gut reaction, or guessing.

>> Utilize more forms of visual thinking and pattern recognition.

>> Process written words and numbers differently than most.

>> Employ perfectionism as a coping response.

And their skills and knowledge may be more likely to be self-taught.

Getting things done

When you plan, organize, or work toward a goal, we call that *executive functioning*. It's how we get things done. Neurodivergent people often do these things differently, for example, by using such tactics as hyperfocus, pattern recognition, or systemic thinking (see Chapter 10 for more). Unfortunately, doing so in a world not designed for them can lead to challenges.

Sensing the world

Whether you're tasting a sour candy or feeling the warmth of a hug, you experience the world through your senses. Neurodivergent people may have stronger or weaker sensory sensitivity. For example, some may be highly sensitive to sounds, lights, or textures, while others don't notice many of them. Most commonly linked with autism, sensory differences are observed in people within other neurotypes, too.

Uncovering Neurodivergent Strengths

What you find challenging and what you find easy is not the same as everyone else. You're not broken if you can't play tennis as well as Serena Williams, nor are you a better person just because you're stronger at something than someone else. That doesn't mean you should shy away from your own strengths. Not at all! (In Chapters 11 and 12 we talk about how to appreciate your own strengths and those of others.)

Here are some strengths commonly associated with neurodivergent people:

>> **Attention:** Attention is the ability to focus on something specific. Those with neurodivergent conditions express attention differently than most. This can include strong pattern recognition and seeing details and pieces of information that others do not readily see.

>> **Focused interests:** One cool thing about many neurodivergent people is their ability to deeply focus on stuff they love. They can enter a flow state, gather loads of information, and generate new ideas. Unlike others, they can work long hours without fatigue. Impressive!

>> **Memory:** Memory is the ability to store, retain, and recall information and past experiences. Our memory includes short-term memory (the brain's temporary workspace) as well as long-term memory (recalling things experienced long ago). Neurodivergent people tend to have a more robust long-term memory when compared to others.

>> **Creativity:** Creativity is the ability to think imaginatively, come up with new ideas, and express ourselves in unique and original ways. Given that

neurodivergent people think differently than most, it enables many of them to offer novel insights, solutions, and ideas.

>> **Empathy:** Empathy is the ability to understand the feelings, experiences, and perspectives of others. Neurodivergent people tend to feel empathy more deeply than others, but they may express their empathy in ways that others are unfamiliar with.

>> **Resilience:** Resilience is the ability to bounce back and recover from difficult situations. Resilience can vary by situation. Each person's experience with it is unique. In many aspects of life, neurodivergent people exercise resilience in ways others cannot.

What is listed here are common patterns of strengths found in neurodivergent people. It is important to remember that neurodivergent people are individuals, and strengths and challenges vary from person to person.

Seeing How Doctors Label Neurotypes

When it comes to how people think, no two human beings are alike. Every brain is unique. Even identical twins — who share a 100 percent match of their DNA — have brains that differ from each other. You may be wondering why we label human brains as belonging to a neurotype if every brain is different. After all, the best way to understand and support an individual is getting to know them as a person. However, humans also love to put people and things into categories so that we can understand them more.

Humans categorize *everything*. There is nothing that we see, touch, or experience that we don't put into one category or another. Don't believe us? Well, you just placed us into the category of *people you don't believe*. Isn't that fun?

Grouping people and things together helps us organize our thoughts and make sense of our world. We group animals into categories of mammals, reptiles, birds, insects, and fish. We separate fruits into melons, citrus, berries, stone fruit, and more. We categorize some things as big and other things as small. We label some items buttons and other objects as forks. You may categorize yourself as happy, sad, healthy, or sick. You may label people who share similar traits as being short, tall, parents, students, teachers, pirates, best friends, athletes, tourists, allergy sufferers, or chocolate lovers.

Medical professionals are human too. They have the same need all of us do to categorize people and things. When doctors sort people into groups based on how their minds and bodies function, these categories are called *diagnostic labels.* The process a doctor uses to determine a diagnostic label is called a *diagnosis.*

REMEMBER

Diagnostic labels are not the only way to understand a person with neurodivergent condition, but they are widely used and can be quite helpful.

Determining what gets labeled what

In general, diagnostic labels are created through a collaborative process involving experts, researchers, and professional organizations. If you've ever worked on a group project at school or at work, it's a lot like that. In creating a diagnostic label, the fact that no two human brains are alike means that there are disagreements, compromises, and heated debates. Grudges are held. Tears are shed. Enemies are created and friends are made. In the end, a workable label is produced — even if not everyone agrees.

A diagnostic label is considered created when it is published in one of several written documents or guides commonly used by the medical community. These are called *diagnostic manuals.* The most common diagnostic manuals include the *Diagnostic Statistical Manual of Mental Disorders (DSM)* produced by the American Psychological Association and the *International Classification of Diseases (ICD)* published by the World Health Organization.

Doctors use any number of diagnostic manuals to diagnose a patient. In the United Kingdom and Europe, doctors may use the *ICD* to categorize an individual as having a neurodivergent condition, while doctors in the United States are more likely to use the *DSM.* Though the categories and diagnostic labels contained within these manuals largely align, they don't always agree.

How diagnostic labels help (or hurt)

You may find it helpful for you or a loved one to be given a diagnostic label. Many do. A label can provide a place to start in better understanding yourself. Plus, being grouped into a category with others like you can help you find acceptance, community, and support. (Check out Chapter 2 for why.)

WARNING

Diagnostic labels have limitations. They don't describe *everything* about a person, and they don't describe every in-and-out and subtle experience of a neurotype. Labels carry the risk of oversimplification, stigmatization, and bias. Some people may wrongly assume things about a person because of a diagnostic label. Others may unfairly use the diagnostic label of a person to marginalize their importance, to dismiss their intellect and talents, or to speak over them. Sometimes, we may use our own diagnostic label as an excuse for doing something we shouldn't.

LABELING NORMAL VARIATIONS AS "DISORDERS"

Have you noticed something similar in many of the diagnostic terms we discuss in this chapter? Autism spectrum *disorder*, attention hyperactivity *disorder*, sensory processing *disorder*, dyslexia, dyspraxia, dyscalculia, and so on. Each of these terms suggests that something is broken or that the individual is deficient — just because they're not like everyone else!

While our modern understanding of conditions such as autism, ADHD, and dyslexia is that they are not diseases or disorders, the language used to describe them often makes it seem that way. One reason why is that doctors rely on diagnostic manuals to understand and talk about these conditions, and the information and language within these manuals tends to be a bit outdated — the *DSM* has only been updated four times since it was first published in 1952, and the *ICD* is updated on average every 12 years.

If doctors use outdated terms such as *disease* or *disorder* to describe neurodivergence, it's largely because that's the language they've been taught to use. How doctors describe neurodivergent conditions will eventually modernize. (It wasn't until 2013 that the *DSM* stopped labeling people who were grieving the death of a parent or spouse as having a "mental disorder.") Give it time.

Why an accurate diagnosis can be hard

You may find that a diagnostic label is a great way to understand yourself or someone you know. (Check out Chapter 10 if you're looking to navigate the diagnostic process.) However, not everyone is able to receive an accurate diagnosis. We call the reasons why *diagnostic barriers*. Here are some:

>> **Complexity:** Neurodivergent conditions can't be diagnosed with a single test. Instead, doctors consider various factors such as medical and family history, as well as observations. This complexity increases the chance that a neurodivergent person may be misdiagnosed.

>> **Time:** Being diagnosed with a neurodivergent condition often requires numerous sessions. Many people do not have the time that is needed.

>> **Cost:** The diagnostic process can be expensive depending upon where you live. While most national health care systems and private insurance plans cover most (or all) of the cost of diagnosing children, this isn't always true when it comes to diagnosing adults.

- **>> Access to diagnostic resources:** Many communities may not have specialists who can diagnose neurodivergent conditions, while in other areas people may face long wait times.

- **>> Limited understanding:** Many doctors have a narrow understanding of neurodivergent conditions due to the training they have received.

- **>> Age group bias:** Significantly fewer doctors are trained in how to diagnose neurodivergent conditions in adults as opposed to children.

- **>> Cultural bias:** Assessment tools are often based on one culture's norms. This leads to misunderstandings for some neurodivergent individuals due to cultural differences. This can cause neurodivergent people to be misunderstood, misdiagnosed, or overlooked.

- **>> Shared traits:** Neurodivergent conditions (such as autism, ADHD, and dysgraphia) share traits, making them "cousins" in diagnosis. Identifying them is challenging for medical professionals due to these shared traits.

- **>> Co-occurring conditions:** Neurodivergent individuals may have additional medical needs. Doctors may find it hard to distinguish neurodivergence from co-occurring conditions.

- **>> Stigma:** The human need to categorize people sometimes results in unfair judgments. This is called *stigma*, and the fear of it stops many from seeking a diagnosis.

It's okay if you are unable to receive a diagnosis because of these barriers. That doesn't make you or your loved one any less neurodivergent.

Moving beyond the diagnosis

You can understand yourself and others in multiple ways. A diagnostic label is just one of them. Anyone receiving a diagnosis of a neurodivergent condition shouldn't stop there. A runner wouldn't quit a race when the official yells "Start!" Likewise, a diagnostic label should be viewed as a starting point of more deeply understanding oneself. It is never, *ever* the finish line.

REMEMBER

Many neurodivergent people are unable to receive an accurate diagnostic label. Does this mean that they have nowhere to start? No, of course not. Even without a formal diagnosis, a neurodivergent person can examine their unique traits, challenges, and experiences to understand who they are.

WARNING

We're not for a moment suggesting that the medical way of looking at differences in brain function is completely useless or not valid. There are many ways of understanding the neurodivergent person, and you may find medical categories to be quite helpful. Many do.

Chapter **6**

Understanding Autism

Often when people ask about neurodiversity, they really mean *autism*. That's understandable; the term *neurodiversity* originated in the autistic community, though it covers much more (check out Chapter 1 for an in-depth look). But just as neurodiversity includes a wide variety of experiences, so does autism. In fact, autism is so wonderfully wide-ranging and beautifully complex that we'd need an entire book to cover it all. So, in this chapter, we focus on the key aspects of autism that are beneficial for everyone to know. Whether you're curious, looking to support someone autistic in your life, or are autistic yourself, this chapter has something for you.

What Is Autism?

Autism is a natural variation in how our brains work, not a defect or disease. It's just a different way of thinking and experiencing the world. Autistic people offer unique perspectives and strengths. By understanding autism as a normal part of human diversity, we can appreciate and support autistic people for who they are. Taking this view helps us see that the world is full of diverse thinkers and learners, each with their own skills and ideas (more on this in Chapter 1). This diversity is vital for a creative and innovative society, reminding us that there's no single right way to think, see, or experience the world.

Every autistic individual is unique, yet each autistic person tends to have differences from non-autistic people in how they think, communicate, socialize, process emotions, and use their senses to explore the world. And in some situations, autistic people's bodies may respond in ways unfamiliar to some. Autism has always been a part of human life for as long as there have been people on this planet. It is just better understood now.

However, the world can be tough for autistic people. It's often designed more for non-autistic ways of experiencing life. This can make things challenging for autistic individuals and their families. Finding understanding and acceptance is hard, and getting services and accommodations needed to make life easier can often be a difficult task.

REMEMBER

Understanding and accepting autistic people makes them feel included and valued. When we celebrate the autistic experience, families with autistic kids benefit too. Acceptance reduces stigma and enables families to feel less isolated and more loved.

WARNING

You'd be wrong to think that the autistic experience is all gloom and doom. It's not. Like anyone, autistic people have their own challenges. But they also experience life in vibrant, wonderful ways. Being autistic has many benefits, and autistic lives are often full of excitement, wonder, and joy (more on that joy in Chapter 19). Autistic people are humans, too.

You may notice an autistic person's differences immediately, or you may know someone for years without realizing they're autistic at all. Despite varied appearances, autistic people share common experiences. We delve into this spectrum of similarities and differences next.

Understanding spectrum conditions

In the previous section, we mention that autistic people have a range of similarities and differences, which is often called the *autism spectrum*. You may have heard this term before. Like many people, you may have first imagined the autism spectrum as a straight line with "a little autistic" at one end and "a lot" at the other. Many people still think of it this way. But the autism spectrum is actually much more varied than that.

When experts talk about autism as a spectrum condition, they're not referring to a simple line from mild to severe. What they mean is that autism includes a wide array of experiences and traits — like a big menu listing lots of things. If you're not autistic, you may see one or two things on that menu that you relate to. But for an autistic person, many things on that menu may resonate with them. Being autistic means relating to a wide range of traits and characteristics listed there. This is similar to how other spectrum conditions, such as attention-deficit/

hyperactivity disorder (ADHD), asthma, and epilepsy, are understood, encompassing a range of experiences and traits.

REMEMBER

Instead of picturing the autism spectrum as a straight line, imagine it as a broad galaxy full of stars. In the autism galaxy, every autistic person is their own constellation (for an illustration of this, see Chapter 1). While all autistic people share similarities, each one has a different combination of traits. Remembering this helps us see people for who they truly are, rather than as a spot on a line.

WARNING

Viewing the autism spectrum as just a line from mild to severe is too simple. It often results in autistic people and their families not getting the help that they need. It's more helpful to see each autistic person as an individual with their own constellation of experiences and traits.

Feeling for the common thread

Being autistic is different for each person. Some autistic people use their mouths to speak, while others communicate in different ways. Some have intellectual disabilities, while others do not. Some need a lot of support daily, while others need just a bit. Each person is unique, and there's no right or wrong way to be autistic. Regardless of these differences, everyone deserves to be treated the same.

"There is no 'typical autistic,'" said Yenn Purkis, coauthor of *The Autistic and Neurodiversity Self Advocacy Handbook* (Jessica Kingsley Publishers). "But I think we probably all like being respected and validated."

Yet, if you put a group of autistic people in a room together, they'd often just "get" each other in a very deep way. There's a common thread woven into every autistic experience, linking all autistic people no matter how different they may look or act on the outside. It's sort of like a shared language or rhythm that resonates with each autistic person.

John from our author team shares: "I'll never forget the first time I talked with an autistic person who used an iPad to communicate. With words being typed into a tablet for me to read, the conversation was slower than I was used to. But soon, we were excitedly chatting about all sorts of things. That first conversation changed how I viewed autism forever. We were both autistic, but we looked, acted, and communicated very differently — yet we still immediately got each other in ways that few could ever see."

Speaking versus nonspeaking

John's story shows something pretty interesting: Autistic people often have lots in common, even if they seem really different on the outside. And just because someone's autism looks different from someone else's, it doesn't mean they're any less autistic. It's all about understanding that autism can show up in many ways, and that's totally okay.

REMEMBER

Even if two autistic people seem very different to you, they may share more similarities than you realize. Every autistic person's experience is genuine and valid, regardless of how it appears to others.

Two groups of autistic individuals who may appear very different to outsiders but actually experience the world in similar ways are *speaking* and *nonspeaking* (or minimally speaking) people.

Around 70 percent of autistic people talk using spoken words to express thoughts and emotions, along with other methods such as facial expressions, body language, writing, and emojis. Folks who use spoken language are often referred to by others as "speaking." This way of talking is similar to how most people talk.

About 30 percent of autistic people don't use spoken words to talk, or may use a very limited number of them. These individuals are often referred to by others as "nonspeaking" or "minimally speaking." Nonspeaking and minimally speaking autistics often communicate using methods different than vocal speech, such as sign language, communication devices, spelling boards, or picture boards — and lots of other ways too.

REMEMBER

Just because someone is nonspeaking doesn't mean that they don't have anything to say. People communicate in many different ways.

Writer Amy Sequenzia, who is autistic and nonspeaking, encourages folks to not overlook the many other ways people can use to communicate too. "We use our eyes, body language, and sometimes, even noises that are hard to understand. People should not only pay attention to these forms of communication, but also support them," she said in a 2012 interview published on the Thinking Person's Guide to Autism website.

WARNING

It's a common misconception that if someone doesn't use spoken words to communicate, they don't have thoughts, aren't aware of their surroundings, or can't hear what others say about them. This is not the case. Assuming this can be very hurtful to the person and can also harm your relationship with them.

"Many people might believe that I cannot think, but despite their thinking, I can," said Neal Katz, who uses assistive devices to communicate instead of spoken words, in a 2012 article published on The Art of Autism website. "What's more is that I listen. A lot of people may stare at me, and when they do, I listen to their body movements and eye gaze. I listen to their ignorance. I listen because I have no choice but to take in the world in the way I can."

How Society Talks About Autism

Understanding the common ways society talks about autism is key to more accurately understanding the autistic experience.

Discarding functioning labels

People often ask whether an autistic person is *high-functioning* or *low-functioning*, but these labels dramatically oversimplify autism, which is an experience far too complex to be neatly categorized. While people with more obvious challenges are often labeled low-functioning, and those with subtler challenges are called high-functioning, in reality, everyone has both challenges and strengths that are often overlooked. Remember, autism is like a galaxy, with each autistic person a unique constellation. Every autistic person has their own blend of strengths and challenges.

As Kat Williams, a mom and autism advocate, said in "The Fallacy of Functioning Labels" published on the National Center for Mental Health website, "'High functioning' doesn't describe how an autistic person experiences autism; it's more about how society sees them."

REMEMBER

Using functioning labels can lead to misunderstandings. It's better to see each autistic person as an individual with their own set of strengths and challenges, rather than trying to force them into a box. Keep in mind that everyone needs some help at times. So instead of getting caught up in labels, let's focus on how we can help each other out.

Autism advocate Laura Tisoncik's famous quote hits home for many: "The difference between high-functioning autism and low-functioning is that high-functioning means your deficits are ignored, and low-functioning means your assets are ignored."

REMEMBER

While we discourage the use of functioning labels, it's important to recognize that some autistic people may need a lot of support, and that's totally okay. Needing help doesn't mean someone is low-functioning. Autism shapes different parts of a person's life in different ways. Some may need a lot of help with communication, daily living, or other things, but the level of help needed doesn't diminish a person's right to be treated with respect.

What about Asperger's?

You may have heard of *Asperger's Syndrome*, which was once a label for autistic people without intellectual disabilities or speech difficulties in early childhood.

But in 2013, experts dropped this term. They realized it wrongly divided autistic people based on a narrow view of autism. Now, autism is understood to include many different experiences. While some autistic individuals may have intellectual disabilities or delayed speech, there's a broad range of other experiences as well.

Honoring how people describe themselves

While the term *Asperger's* is no longer widely used, some may still choose this label. That's okay. It may be their original diagnosis or they have other reasons for it. Everyone's self-acceptance journey is unique. Some may say "I'm autistic," others may say that they are a "person with autism," or some may still use "Asperger's" to describe themselves. It's about what feels right to them.

TIP

If you're not autistic, it's generally better to say "autistic person" rather than "person with autism." Research shows that's the preference of most autistic people. You may have been taught to say "person with autism," particularly if you work in education or health care. But keep in mind, this advice probably didn't come from an autistic person.

But don't stress too much about the perfect words when discussing autism. Shannon Des Roches Rosa, senior editor and cofounder of the Thinking Person's Guide to Autism website, puts it well: "It's more about the intention behind your words, not the exact words you use." Being genuine and respectful is key.

How the Autistic Brain Thinks

The autistic brain is truly remarkable. It functions differently from most people's brains, and that's not just okay; it's fantastic. You see, the world has always needed a variety of thinkers (more on that in Chapter 3), and autistic people have always played their part.

Understanding how the autistic brain thinks benefits everyone. For autistic folks, knowing how your brain works is like having a map to navigate life more effectively. And for everyone else, understanding autistic brain thinking helps you appreciate and support the autistic people you know.

Patterns of thinking

Our brains are incredible and offer many different ways of thinking. For example, critical thinking is used to analyze and evaluate information; logical thinking follows clear, rational reasoning; and emotional thinking bases decisions on feelings or our "gut instinct." No single way of thinking is better than another; each plays an important role in our thought process.

Linear versus nonlinear thinking

Humans most frequently use two additional thinking styles: *linear* and *nonlinear*. Linear thinking follows a straight line, moving from one point to the next. It's ideal for solving problems in a straightforward manner, similar to baking a cake where you plan, measure, and follow each step in order.

Nonlinear thinking looks more like a web. It jumps around, forming creative and unexpected connections. This type of thinking is great for creative tasks or situations where out-of-the-box thinking is needed. Imagine nonlinear thinking as planning an adventurous road trip where instead of just following GPS directions, you explore different routes, consider the fastest versus the most scenic paths, and choose the best places for stops, adapting as new shortcuts are found.

Different neurotypes have different preferences

In everyday life, most people primarily use linear and emotional thinking. Linear thinking is useful for routine tasks, such as planning your day. Emotional thinking is important in social situations and making quick decisions.

Autistic brains work a bit differently. They often switch between a lot of nonlinear thinking (the web mentioned earlier) for planning and logical thinking to keep things on track. Autistic brains can also use linear and emotional thinking when necessary, but these aren't usually their go-to methods. When autistic people use their preferred thinking styles, they often thrive. However, fitting their unique thinking style into a world that mainly values linear and emotional thinking can be challenging. For example, an autistic person may excel at creatively solving a complex problem, but struggle with making a decision that requires a quick response.

REMEMBER

Both linear and nonlinear thinking have their advantages. Linear thinking offers a straightforward approach to problems, while nonlinear thinking leads to creative solutions. And it's important to note: Autistic people can use both linear and nonlinear thinking, just as non-autistic people do. They may just have a preference for one over the other.

Elementary, my dear Watson: Associative thinking

Nonlinear thinking often involves what is often called *associative thinking*. Think back to the web of nonlinear thinking discussed earlier. Associative thinking is about seeing how different ideas and pieces of information in this web connect. Autistic people are often skilled at this, much like the detectives in mystery novels (in Sir Arthur Conan Doyle's series of books, Sherlock Holmes often is seen complaining to his assistant, Watson, about the lack of associative thinking skills among the local police).

In daily life, autistic people use associative thinking to understand their surroundings. This makes them adept at noticing patterns and connections that others may overlook. Such a skill is valuable for problem-solving, learning, and grasping complex systems. They may realize how a certain sound indicates it's about to rain, or link a recent conversation to one from months ago. They may identify trends in data that others overlook or recall detailed facts about a topic and see how it impacts something seemingly unrelated.

Cloud of thoughts: Autistic daydreaming

Autistic people often engage associative thinking through daydreaming, letting their minds wander from their surroundings into a realm of nonlinear thinking. They use these daydreams for solving complex problems, learning new things, playing, or simply thinking about things they enjoy. This is so prevalent that autistic children are often said to be "lost in their own little world." However, they're not lost, and this trait isn't just confined to childhood. Their minds are actively growing, learning, and thinking up new things.

It makes perfect sense: Using logical thinking

Autistic people are often quite skilled at figuring things out through logical thinking. It's like building a structured framework where everything fits together in an organized manner. In everyday life, this means they can plan things meticulously. For example, they may carefully allocate specific times for various activities to ensure smooth and efficient routines.

When solving problems, autistic individuals often analyze issues methodically. It's similar to troubleshooting a computer by systematically checking each part. This approach is helpful in many areas of life, bringing calm, order, and efficiency.

REMEMBER

While logical thinking and nonlinear thinking may seem different, they actually work well together. Autistic people use logical thinking for structured tasks such as planning and problem-solving, while nonlinear thinking aids in creativity and discovering new ideas.

Focused interests

Intense interests are a key part of the autistic experience. These are topics, activities, or hobbies that the autistic person can develop deep passion and knowledge about. These interests can bring immense joy, and thanks in large part to associative thinking (and hyperfocus, which we discuss later in this chapter), autistic individuals can become experts in these areas faster than most. And autistic-focused interests often lead to incredible insights, discoveries, and creations, benefiting not just the individual but the whole world.

As a child in Canada, actor Dan Aykroyd had a strong fascination with the paranormal. He dedicated himself to learning everything he could about the subject, nurturing his knowledge over the years. "I became obsessed by Hans Holzer, the greatest ghost hunter ever," said Aykroyd, who is autistic, in the 2013 article, "I Have Asperger's; One of My Symptoms Included Being Obsessed with Ghosts" for *The Daily Mail*. "That's when the idea of my film *Ghostbusters* was born."

REMEMBER

Focused interests for autistic people are as unique as fingerprints. Some may have a single passion, while others enjoy a variety of interests driven by their curiosity. These interests can change and grow as they do. Like everyone else, autistic individuals have diverse interests that shape their identity.

How the Autistic Brain Organizes Thoughts

Humans use what's known as *executive functioning* to do all the little things that help us get through the day. Think of executive functioning as the brain's captain, helping us plan, organize, and finish tasks, including managing time, focusing, and making decisions. Everyone uses these skills differently, influenced by such factors as neurotype, culture, and experiences. Some may be great at planning but poor in impulse control, while others may multitask well but struggle with time management.

Different cultures can have preferences for certain types of executive functioning skills, and this can affect how easy or difficult it is for linear or nonlinear thinkers to thrive. In some cultures, executive functioning skills such as strict time management and detailed planning are really valued. This can be great for linear thinkers, who naturally think in a step-by-step structure. But for nonlinear thinkers, like many autistic folks, it can be tough.

REMEMBER

While sometimes seen as having executive functioning challenges, people with different thinking styles can excel in environments that value their unique approach. Autistic individuals, for example, can show high determination and focus when their logical thinking aligns with planning and task management.

The use of routines

Autistic individuals often use routines to organize their thoughts, finding comfort and predictability in them. These routines help them as nonlinear thinkers by providing structure and familiarity, easing navigation in their world. Moreover, routines can foster linear thinking, merging it with broader thought patterns to better understand their surroundings and gain control.

Routines for autistic people can take many forms. They may involve following a daily schedule, always showing up for an activity they signed up for, playing the same rotation of games every day to center their thinking, using checklists, sticking to a set route to work, consistently choosing the same library seat, or maintaining a specific order for things. While some routines are clearly visible, others are so subtle they may go unnoticed — even by the autistic person themselves.

It's important to note the autistic person may use routines to help them with other parts of life, such as managing sensory sensitivities or feeling secure. But when applied to thoughts, routines can act as a thinking funnel (see Figure 6-1). This can help the autistic person channel their expansive, nonlinear thinking into focused, concrete thoughts or actions that are easy to follow. It's a way to streamline their unique thought processes into practical, everyday applications.

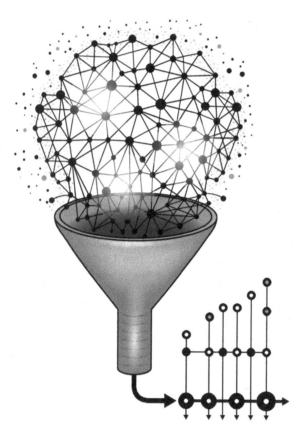

FIGURE 6-1:
Nonlinear
thinking funneled
down to linear
thinking
visualized.

Disruptions in routine

Disrupting routines can be jarring, especially if the autistic person is using them to organize their thoughts. It's like having the map they rely on for their nonlinear thinking suddenly taken away. This can make thinking and communicating more difficult, increasing stress.

REMEMBER

Some may call this inflexibility, but that's not quite right. Autistic people can be flexible, just like anyone else. It's crucial to understand that routines provide predictability and comfort, making daily tasks easier for autistic individuals. When changes occur, they can adapt, but they often require time and support to adjust. Quick changes can be disorienting or uncomfortable for them. They can be flexible, but adapting to new situations or changes in routines may take some time and help.

How the Autistic Brain Uses Focus

Autistic people also use focus to organize their thoughts and think up new ideas. And when an autistic person is motivated to focus on something, they can really zoom in. It's a trait often shared with their "autistic cousins" — people with ADHD.

Entering the flow state

This high level of focus is called *hyperfocus*, where a person can concentrate on a topic or task without distraction, entering what's known as a *flow state* (we discuss this more deeply in Chapter 7). This intense concentration is a strength, allowing autistic people to excel in areas they're passionate about. But it can be challenging too, as it's easy to forget other things that need to get done. In addition, shifting out of hyperfocus can be hard.

REMEMBER

Focused interests discussed earlier is different from hyperfocus. An autistic person may use hyperfocus on a focused interest, but they're not the same thing.

Transitioning out of hyperfocus

Transitioning out of hyperfocus can be tough. It's like having a movie suddenly turned off — a bit disorienting. Autistic people need time to ease out of hyperfocus. Forcing a quick switch can be jarring, like slamming on the brakes while driving fast.

TIP

A gentle warning before changing tasks and allowing the person to reach a natural stopping point, helps the transition from hyperfocus.

How the Autistic Brain Uses Memory

Autistic people often have a unique and interesting way of remembering things. Many have a powerful long-term memory, able to recall details that others may forget. However, their short-term memory, like remembering recent instructions, may require more effort. It's as though they have a huge library for older memories and just a small notebook for newer ones. The library has plenty of space, but the notebook can only fit a few things at a time.

How the Autistic Brain Uses Senses

Taste, touch, sight, sound, and smell — our five senses are so important that they're one of the first things we're taught in school. They're the foundations for how we understand ourselves and explore the world.

Imagine that your senses were louder, bolder, brighter. Picture what it would be like if you could feel texture in incredible detail, hear slight differences in sound from three blocks away, or taste vivid flavors hidden in foods that others find bland. Sounds amazing, right? But what if this heightened sensory experience changes unexpectedly? The same sensations that bring you joy in one moment may become overwhelming in another. They may cause pain, make it hard to focus, difficult to speak, or make you feel ill.

Autistic people often experience the world through their senses in a unique way. It's like they have a special set of antennas picking up more signals than the average person. This can be pretty cool in some ways. For example, they may enjoy the beauty in small details that others overlook, or they may have a great ear for music or an excellent taste for flavors.

But this heightened sensory sensitivity comes with its own set of challenges. For autistic people, everyday sounds such as traffic or a noisy crowd may be too loud or distracting. They may find strong smells or certain textures really uncomfortable. While their unique way of sensing the world can be a plus, they can also make everyday situations, such as hanging out with friends, going to school, or working, tougher. In these environments, distractions such as loud noises, bright lights, or large groups can be overwhelming, making it difficult for them to concentrate or feel at ease like others.

Some autistic people may be less sensitive to stimulus such as pain, loud noises, certain smells, or physical touch. This may lead to seeking more powerful sensations such as very loud music, intense flavors, or strong physical pressure. Beyond the five overt senses, autistic people may experience three subtle senses differently: interoception (feelings internal to your body such as hunger), proprioception (knowing where your body parts are such as your hand's position), and spatial orientation (understanding where you are in a room or space). These senses help you move and know your body's needs.

REMEMBER

Every autistic person has their own unique way of experiencing their senses. What feels good to one person may be painful for another. Also, whether the sensory perception is a challenge or a strength can vary from situation to situation for each person.

How the Autistic Brain Processes Emotions and Exercises Empathy

Autistic people often show their emotions differently. Like those with ADHD (discussed in Chapter 7), they may feel emotions more intensely. This leads to strong joy and empathy, adding depth to their emotional experiences. However, they experience tough emotions such as frustration or sadness more deeply as well. Managing these intense feelings can be challenging. It's essential to provide support, understanding, and patience to help them process and express their emotions.

Autistic people may express empathy and emotions in ways unfamiliar to others, but this doesn't mean they don't feel them. Many have a deep sense of empathy, often more profound than in neurotypical individuals. They also frequently have a strong sense of justice and fairness, guiding them toward doing what's right. They may express empathy by speaking honestly and directly, tackling tough topics head-on and openly showing emotions such as joy and distress. When responding to others' problems, they often focus intensely on finding solutions and may ask many probing questions along the way. While this may seem blunt or unemotional to some, it's simply their unique way of expressing a deep level of empathy.

How the Autistic Brain Communicates

On a Saturday morning in San Francisco, several dozen autistic adults sat in silence, quietly sipping their coffee inside a community center meeting room somewhere south of Market Street. That was until a gruff-throated voice began to scream.

"DOES DAVID BYRNE KNOW HE HAS AUTISM?! DOES HE KNOW?! IF NOT, SHOULD WE TELL HIM?!"

As soon as the question was yelled, dozens of conversations began. Phones were brought out, websites searched. Points and counterpoints were shouted back and forth. Within minutes, the noise began to dampen as it was confirmed and then verified twice over: David Byrne, the lead singer of The Talking Heads, is autistic and well-aware of that fact.

Autistic people often use communication to express themselves in ways that differ from the non-autistic people around them. "When you have trouble expressing yourself through the normal channels, you find other ways to do that," said Byrne, telling the BBC in 2023 that his difficulty navigating neurotypical communication styles pushed him toward the stage at an early age.

For an Emmy-, Tony-, and Academy Award-winning artist like Byrne, differences in how they communicate may mean using carefully rehearsed scripts on and off stage. They can also mean shouting out exactly what is on your mind, much like the person in the story earlier. If you're not autistic, you may be unfamiliar with common ways autistic people communicate. We go over a few of them here.

Using spoken language

There's no single correct way for people to speak. The way people talk varies and is often shaped by their culture. Some cultures prefer formal speech, while others are more casual. In some places, people learn to speak very directly, while others are taught to express themselves in less straightforward ways.

Let's be clear: Understanding direct versus indirect language

When people say exactly what they mean, this is called using *direct language*. On the other hand, when they hint at their meaning or use suggestions instead of saying it outright, that's known as using *indirect language*. All people and cultures use a mixture of both, but usually one style is more dominant.

REMEMBER

Autistic brains often prefer direct language. "We think it, we feel it, we express it; whether we've known you for 20 seconds or 20 years," noted Jamie A. Heidel in her article, "Why Small Talk is Important for Neurotypicals (An Autistic Perspective)" for *The Articulate Autistic*. For autistic individuals, this directness isn't about rudeness; it's about clarity and honesty. However, people who speak indirectly may see this as too blunt, particularly if they're accustomed to more subtle communication styles.

Due to a preference for direct language, an autistic person may find it easier to understand communication in cultures that prioritize clear and direct talk (such as Germany) versus places where people speak in less direct ways (such as the United Kingdom). For example, take a look at how two cultures might say the same thing:

> **United Kingdom:** I think there might be a few areas of your report that could use a bit more attention. Perhaps you could take another look?

> **Germany:** Your report has several errors. You should revise it before the meeting.

The second example is easier to understand for some who like to speak directly. Of course, the example we use is a broad generalization. So don't interpret it too literally.

Say what you mean: Interpreting language literally

When people speak directly, they usually expect others to do the same. This means that if a direct speaker hears someone using indirect language — such as hints or sarcasm — they may take it exactly as it's said.

Take Karen and Noah, for example. They're cooking dinner, and Karen feels cold. She says, "Wow. It sure is chilly in here," hinting she wants Noah to close the window next to him. But Noah, thinking directly, just agrees, replying, "Yeah, it is pretty cold," and keeps cooking away. Karen was indirectly asking for the window to be closed, but Noah took it as a simple statement, not a request.

This frequently happens with autistic people. In the past, people believed it was a problem with the autistic person. It's actually because people use language in different ways.

TIP

If you want to get better at noticing when someone is using a different speaking style, you first need to understand and accept your own. "I often speak less bluntly with non-autistic colleagues and friends," says John from our author team. "And they try to speak more directly with me. None of us are changing who we are, but all of us are happy to give each other the accommodations we need."

Get a clue: Hinting at things

In the earlier example, Noah interpreted Karen's statement about it being chilly quite literally. Karen, on the other hand, was using a common method of indirect communication: giving hints and suggestions. People who speak indirectly often expect the listener to pick up on these subtle hints and react appropriately. They may choose to speak this way to be polite, to avoid seeming aggressive, or to make the other person feel more comfortable.

That's just great: Using sarcasm in speech

Sarcasm is a form of indirect language commonly used to convey disapproval or criticism, often in a humorous way. It can serve as a coping mechanism in uncomfortable situations, used as a playful means of bonding, or simply to be funny. It's a myth that autistic people can't understand sarcasm at all, but depending on the person and the situation, they still may have difficulty with it. "I typically get when friends are being sarcastic," says John. "But a small part of me still wonders if they're being serious or not."

Give 'em something to talk about: Using small talk

Indirect speakers often like to use small talk as a warm-up for conversations. Chatting about everyday things such as sports scores or traffic helps them relax, socially connect, and set the stage for deeper discussion. To say that autistic people find small talk pointless and insufferable would be exaggerating — *slightly*. The truth is that many autistic people much prefer deep, meaningful conversations over casual chitchat. While discussing the weather may not be their thing, if you need to confess your deepest problems to a perfect stranger, an autistic person may be a great choice.

While different preferences for small talk may not seem significant, choosing not to engage in small talk can have downsides in cultures that value it. Not participating in small talk can result in missed opportunities for social bonding and networking, and may be seen as unfriendly.

"It's also kind of a huge barrier to getting employed because job interviews rely way more on your perceived 'social skills' than on your ability to actually do the job," said autistic actor and composer Syndey Zarlengo in her YouTube video titled "Small Talk — An Autie's Guide to Neurotypical Social Skills." "So, getting good at small talk could very much be the difference between getting a job and not getting a job." Zarlengo encourages autistic folks to embrace their own speaking style but to also use small talk as an accommodation for those who may need it. "People have more in common than you think and it's really fun trying to find those little connecting pieces."

Using body language

Just as the way people speak can vary from culture to culture, body language — including facial expressions, gestures, and our posture while talking — also changes from place to place. For example, a gesture or facial expression that's friendly in one culture may be interpreted as rude in another. This means the way people use their bodies to communicate can have different meanings depending on where they are, or on the stereotypes and biases people hold.

"As a Black woman living in the United States, I am always mindful of what others think about me, and the assumptions they may make," said Catina Burkett, a social worker in South Carolina in the *Spectrum* magazine article, "Autistic While Black: How Autism Amplifies Stereotypes." "As a Black woman with autism, I am especially aware that colleagues often see me as an 'angry black woman,' even though my thoughts and behaviors are the opposite of this stereotype."

These differences can also mean different things depending on our neurotype. Autistic people often understand each other's facial expressions, gestures, and body language quite well, just as non-autistic people do among themselves. However, when autistic and non-autistic people get together . . . well, things tend to fall apart.

REMEMBER

Autistic and non-autistic people often aren't taught to understand each other's body language. Just as there are differences in speech, it's important to acknowledge that body language may be interpreted differently too.

Body positioning

Non-autistic people often rely on body language to indicate interest in a conversation or to aid their focus, such as leaning forward to show engagement or crossing arms when feeling defensive. Conversely, autistic individuals can engage deeply in conversations without depending on such body language cues.

TIP

Autistic individuals often prefer to sit side by side when talking with a person. If you need to have a deep or difficult conversation, follow the autistic lead and sit next to the person instead of facing them. Research shows this relaxes both people and encourages openness. Experts call this the "side-by-side effect," but we like to call it "the autistic approach."

Eye contact

Non-autistic individuals often use eye contact in conversations to connect, show they're listening, or express emotions. However, for some autistic people, maintaining eye contact can be overwhelming or distracting, hindering their focus on the conversation. In addition, it's common for autistic individuals to feel uncomfortable or even physically pained when making eye contact.

"Avoiding eye contact is one of the things I find myself automatically doing to minimize the quantity of incoming sensory information," said Judy Endow, an autistic clinical worker in Wisconsin in her blog post titled "Eye Contact and Autistic Dissociation." "I don't consciously think about avoiding contact. It just happens because that is how my brain works."

Never force eye contact when interacting with an autistic person. Instead, mirror the way they are using eye contact. Lack of eye contact is not a reflection of their level of interest or attention. They are just communicating in a way that feels natural and comfortable for them.

Vocal volume and tone

Autistic individuals may also have unique ways of physically speaking. Some may naturally speak louder than average, while others may speak very softly, making it hard for people to hear them. Some autistic people may prefer to speak in short, one-word sentences. Others may talk on and on. In addition, some autistic people may use a flat tone that can sound unfamiliar to others. However, all these ways of speaking are valid.

How the Autistic Brain Socializes

A while back, John from our author team pulled a close friend to the side and confessed that he hadn't been asked out in years. "It's so frustrating. I get invited to dinner, movies, concerts, picnics, plays, but never on a date!"

"John, those were dates!" his friend exclaimed. "What do you think 'being asked out' means?" It turns out John had been waiting for someone, *anyone,* to explicitly ask, "Would you like to go out on a date?" He was surprised to learn that most people imply that it's a date without using those exact words.

"You mean I dated *all* those people without even realizing it?" asked John.

Yes. Yes, he had.

Autistic and non-autistic people often socialize in different ways. This can include differences in how people socially communicate their intent, as John's story shows. Other differences include how people bond, participate in activities, and show support for one another.

Making and maintaining friends

Autistic people's approach to forming and maintaining friendships can differ from the norm. This doesn't imply their methods are better or worse than others, but they are different and it's helpful to understand how.

Bonding through shared interests

Autistic people often connect with others by sharing their interests or showing genuine interest in what's important to someone else. For example, if an autistic person loves painting and meets someone who also enjoys art, they can bond by talking about painting techniques, favorite artists, or even creating art together. This shared interest in art becomes a foundation for their relationship, allowing them to connect deeply through their mutual love for creativity. Whatever the hobby or topic, sharing interests can create deep bonds between autistic and non-autistic people.

Being faithful and loyal friends

Autistic people are often seen as dependable and trustworthy friends, and there are good reasons for this. Many autistic individuals highly value honesty and consistency, which makes them trustworthy and reliable. If an autistic person says that they'll do something, they'll most likely do it. And their straightforward way of communicating means they're less likely to engage in deceptive behavior or play games, adding a layer of reassurance in friendships. As a result, friendships with them can be incredibly fulfilling.

Barriers to making friends

Autistic people often have a tough time making friends, but it's not their fault. They may struggle to understand social cues and unwritten rules that are used by non-autistic people. Figuring out when to start or end a conversation, or what to say in various situations, can be difficult. They may also get overwhelmed in busy or noisy social environments, and they may worry about whether they're meant to be included or not. However, these challenges don't mean they aren't interested in making friends. It's just that neurotypical socializing may not always be ideal for them, and they may need different ways of connecting that suit their unique needs and preferences.

Challenges in feeling included

One reason autistic people may struggle with making friends is that they often need clear signals to feel welcome in social situations. They may not pick up on the subtle hints that non-autistic people use, so without a straightforward invitation, they may hold back or stay quiet. They're also likely to not join in on social activities unless they're explicitly asked. Being direct and clear can really help. John jokes, "Autistic people are like vampires on your doorstep; we both need a clear invite to come in."

REMEMBER

When folks are understanding and accepting, it's much easier for autistic people to make friends with them.

Needing alone time

Autistic individuals often require alone time, which means having time by themselves without others around. This is essential for them to recharge and feel at ease. During these moments of solitude, they can immerse themselves in their focused or intense interests, unwind, or simply enjoy some peaceful moments. It's not about avoiding others but rather about striking a balance between social interactions and personal space to ensure their well-being. For the autistic person, alone time acts as a reset button.

Navigating additional challenges

Everyone has times when socializing is difficult or awkward, and autistic people are no exception. But the specific challenges autistic people face in social interactions are often overlooked.

Abuse of trust

Autistic individuals are often known for their high level of trust, something that their friends come to love. However, this trust also has its downsides. It can cause misunderstandings, lead to pain when someone else doesn't follow through on their word, or leave autistic people more open to exploitation. As a result, some may become cautious or hesitant to form new relationships in order to protect themselves. Finding the right balance between trust and caution is a tricky part of building and maintaining social relationships.

Not remembering faces

Some autistic individuals struggle with prosopagnosia or "face blindness," finding it hard to remember faces. They may not recognize someone they've met several times. While about 2 percent of the general population experiences face blindness, studies show that 20 percent or more of autistic individuals may experience it to varying degrees.

The experience of face blindness varies, with context playing a crucial role. For example, someone with face blindness may meet a couple at a dinner party and enjoy a conversation. But if they encounter one of them alone on the street the next day, they may struggle to recall where they met them before. Without the dinner party setting and the other person's presence, recognizing the face becomes more difficult.

TIP

John offers this tip to anyone else who has this trait: "Follow the lead of celebrities and politicians who often greet people by saying 'Nice to see you!' instead of 'Nice to meet you!' It's a trick they use since they meet so many people, they're never quite sure who they've been introduced to before."

How Autistic Bodies Respond

Our bodies often respond to the world around us through feelings such as stress, needing to rest, or needing to move. Autistic people experience these needs too, but their bodies may express themselves differently.

Stimming

The next time you go to a concert, look around. As soon as the music begins, you'll see people standing, raising their hands, swaying their bodies back and forth. This is called *stimming* (short for self-stimulation), and it's a natural part of life.

Understanding why people stim

Stimming is something all humans do — this includes you! It's repetitive movements or sounds that help us deal with emotions, stress, sensory input, or focus. Tapping your foot, fiddling with a pen, pacing back and forth, and twirling your hair are all examples of stimming. So is doodling, dancing to your favorite song, and humming a tune.

People used to think that autism caused stimming, but that isn't true. Every person needs to stim, and it's just that some people need to stim more than others. This includes autistic people, but it also includes other folks, such as those with ADHD or anyone with tickets to a Beyoncé concert. (Seriously, who can sit still at a Beyoncé show?) Whether it's Beyoncé, Taylor Swift, or your favorite band, life offers many sensory and emotional experiences that may make you want to stim. So go for it!

REMEMBER

It's important not to stop an autistic person from stimming because stimming helps them regulate their emotions, cope with stress, and process sensory information. It's a natural and healthy way for them to manage their feelings and stay focused. They need to stim — and so do you.

Understanding differences between stimming and SIB

Some people may engage in what's known as self-injurious behavior (SIB), which describes actions an individual does to intentionally harm their own body, such as hitting, biting, or scratching themselves. This includes some autistic people. SIB is usually a sign of distress, discomfort, or an attempt to communicate something when other forms of communication may be limited. While both stimming and self-injurious behavior involve repetitive actions, and may sometimes intersect, at their root they are two different things.

Understanding masking and autism

Autistic masking is when people hide their true selves to mimic non-autistic behavior. They do this to blend in and avoid negative reactions. However, constantly monitoring and altering behavior can be exhausting and stressful.

Fear and worry often lead autistic people to mask, even if they don't want to. "My fear is that if I don't mask, push through, and show how capable I am, I won't be offered opportunities in the future or be valued the same," said Emily Swiatek, an employment consultant in Liverpool, in the article, "Autism at Work: Achieving Inclusion for a Spectrum of Needs" published on the Headstart website.

Many autistic people use masking for short-term gains, such as not having to explain your autistic traits at a party or not drawing attention to yourself when you just want to buy a cup of coffee at the corner shop. Some find that masking can be useful in a world that doesn't always accept neurodivergent traits. However, masking takes a lot of energy and effort and can have dramatic and negative long-term effects.

Musician Leah Reinardy explains that she first used masking to get by in school. In her 2022 TEDx Talk titled "Autistic Masking: A Dangerous Survival Mechanism," she said, "The more teachers saw me as 'normal' the higher my grades would be," and added, "Hell, if I put on a strong enough social performance, I might even make a friend." But constant masking is tiring because it requires effort and always being on guard. Over time, this can lead to feeling alone, losing your sense of self, depression, and autistic burnout. Reinardy also shared, "It came with a price. I was in therapy by age eight and on antidepressants by ten. By the time I hit middle school, I was this model student by day but would come home and completely break down at night."

REMEMBER

Autistic people sometimes hide their true selves because of how others may react. We can help by getting to know them, accepting their differences, and creating spaces where they feel safe to be themselves without stress.

The autistic body under stress

Picture being in a place where lights are overly bright, sounds are too loud, and sensations are intense. This is often the everyday experience for many autistic individuals. In addition, social interactions can feel like constantly learning a new language, increasing their stress. Consequently, daily life can be quite stressful for autistic people. In the following sections, we explore how their bodies sometimes react to such extreme stress.

Understanding autistic meltdowns

Autistic meltdowns are intense responses to overwhelming situations, similar to a cup overflowing from too much water. They can be triggered by such things as loud noises, bright lights, stress, routine changes, or social pressures.

An autistic person may sense a meltdown approaching, with feelings of being trapped, building internal pressure, or by increased sensory sensitivity.

REMEMBER

During an autistic meltdown, the overwhelming emotion isn't anger but an intense blend of fear, anxiety, and stress that few others can truly comprehend. Meltdowns may look like anger because autistic individuals may cry, shout, bang their bodies against objects, or throw things. They may also struggle to talk or communicate normally. During a meltdown, autistic people may use coping mechanisms such as stimming, fleeing, hiding, or squeezing their bodies into small spaces.

Meltdowns aren't tantrums, but the body attempting to deal with a mounting wave of stress. Meltdowns may cause panic — a fierce tide of fear that causes physical reactions such as a pounding heart, shortness of breath, or dizziness. Alongside these physical symptoms, people in the throes of a meltdown often feel a profound sense of helplessness — like they are sunk into a deep hole with no way to climb out.

The length of meltdowns varies greatly and depends on the person's coping methods, their environment, and the specific circumstances. They can last a few minutes to over an hour, ending either slowly or suddenly.

TIP

If an autistic person near you is having a meltdown, stay calm and create a safe, quiet space for them. Give them some physical space, speak softly, and be patient. Offer support but avoid giving advice. Applying deep pressure, such as with a weighted blanket or a strong-but-gentle hug can also help.

WARNING

Deep pressure can bring relief, but always ask for permission. Respect their consent because unwanted physical contact, even well-intentioned, can increase anxiety or discomfort during a meltdown.

After a meltdown, autistic individuals are often exhausted, both emotionally and physically. Many autistic people have compared it to being "hit by a truck" or "run over by a train." Embarrassment or shame may follow, making them uncertain about returning to social situations or interacting with those who witnessed the meltdown. Finally, they may feel relief.

TIP

If you're a parent, you can make a difference by explaining meltdowns to your autistic child. Offer reassurances such as, "I'll always understand, accept, and love you," and work with your kid to create a "meltdown plan" about what to do when meltdowns happen. Just make sure you do all this during calm times, spaced far away from a meltdown.

Understanding autistic shutdowns

Autistic shutdowns may be compared to a computer going into sleep mode. When stressed or overwhelmed, an autistic person may shut down, becoming quiet, withdrawn, or unresponsive. They may have trouble speaking, walking, or moving. It's the brain's way to handle too much information or stress.

Shutdowns differ from meltdowns. In a meltdown, stress may cause visible reactions. Shutdowns are more about feeling tired, disconnected, or shutting off from the world. The person may feel worried, embarrassed, or even calm. Shutdowns can happen quickly or slowly and last from minutes to hours. Recovering from them takes time and often leaves the person exhausted. Giving them space and patience during recovery is important.

REMEMBER

Remember, someone in a shutdown isn't ignoring you; their brain is just saying "I need a break."

Understanding autistic burnout

Autistic burnout is similar to a battery running out of juice after too much use. It happens when an autistic person deals with too much stress or demands for a long time. During burnout, they often feel very tired and may find even simple tasks hard. They may lose interest in activities they used to like, and everyday tasks can become overwhelming. They may feel as though they have no energy or motivation left, and their senses may get more easily overwhelmed.

"I'm not typically an irritable person, but during burnout, even the slightest sensory trigger can completely overwhelm me," said Megan Ana Neff, a clinical psychologist who is also autistic, in a 2023 article in *Neurodivergent Insights*. "It's as if my sensory threshold has become incredibly thin and delicate."

Recovering from burnout can be a slow, gradual process. It requires plenty of rest, less stress, and sometimes changing things in life to make challenges easier to handle. It's really important for the person to move at a speed that's comfortable for them. Many autistic people find relief from burnout by cutting down on things that overstimulate their senses or demand too much socially. Getting advice from professionals or following a self-care plan that works for them can also help.

REMEMBER

After burnout, an autistic person may be more careful about how they use their energy, changing routines or habits to avoid getting overwhelmed again.

Understanding autoimmune and GI issues

Many autistic individuals experience autoimmune or gastrointestinal (GI) problems. The exact causes are uncertain, with genetics, sensory sensitivities, and stress being possible factors. Research into the connection is ongoing. Addressing these issues is crucial for the well-being of autistic individuals. It's recommended to seek personalized advice from health care professionals.

Understanding anxiety, trauma, and PTSD

Being autistic in a world designed for non-autistic people brings a lot of stress to the autistic person. Often attached to this is anxiety. As a result, autistic people often experience anxiety disorders. In addition, and for various reasons, autistic people may also experience trauma and post-traumatic stress disorder (PTSD) (more on this in Chapter 12). With proper support and strategies, many discover ways to cope and thrive.

Empowering Autistic People

Empowering autistic individuals involves giving them the necessary tools, support, and opportunities to make their own choices and manage their lives. It's about acknowledging their capabilities, hearing their voices, and respecting their choices. This empowerment allows them to lead fulfilling lives and contribute their unique talents to the world.

Shifting our perspective

Society needs to view autistic people not as flawed versions of others, but as full versions of themselves. Autistic people possess distinct qualities including original perspectives, intense passions, and a unique worldview. Empowering them involves creating environments in schools, workplaces, and homes that celebrate these strengths. This means developing sensory-friendly spaces, respecting their ideas, and ensuring clear communication. Providing positive support and encouragement is also key (for a look at some practical steps to take, check out Chapter 24).

Taking strengths and challenges seriously

To support autistic people effectively, it's crucial to recognize both their strengths and challenges. Sometimes, people may not believe the difficulties faced by those who don't obviously show their autism, or society may overlook the talents of those who need extra support. It's important to see each autistic person as a complete individual, with both ups and downs.

Many autistic people were never diagnosed as children and encounter significant obstacles when seeking an assessment as adults (see Chapter 5). Regardless of diagnosis, if a person says that they're autistic, believe them.

Encouraging self-advocacy

Self-advocacy is crucial for autistic individuals. It involves them speaking up for themselves and expressing their boundaries, strengths, and needs. Because they know themselves better than anyone, self-advocacy helps them to secure support and accommodations which work best for them (to learn more about self-advocacy, turn to Chapter 13). It's about being their own champion, ensuring that others understand their perspectives.

Deciding on therapy

When it comes to therapy for autistic individuals, it's crucial to adopt a neurodiversity-informed approach. This means understanding and respecting that different neurological conditions are natural variations of the human brain, not defects to be fixed. The goal of any therapy or service should not be to change the autistic person, but to support and embrace their unique way of doing things so that they can learn and develop strategies to live life on their own terms and thrive in their own way.

Closely Associated Conditions

The autistic experience is complex, and a number of conditions are closely associated with the neurotype, including intellectual disabilities, epilepsy, and sleep differences.

Conditions similar to Autism

Four conditions share many characteristics of autism. In fact, they share so many similarities that a growing number of experts believe they are autism interpreted in a different way. Some argue that those diagnosing these conditions often miss or misinterpret other autistic traits.

Nonverbal learning disorder (NLD)

Nonverbal learning disorder (NLD) is a label sometimes applied to a person who is said to struggle with interpreting facial expressions and body language, motor skills, and judging distance and space. Comedian Chris Rock was diagnosed with NLD after friends suggested he get assessed for autism. Rock had long recognized his difficulties in socializing and communicating with others. "I just always chalked it up to being famous," he said in an interview on *The Today Show* in 2020, thinking it was others always acting nervously or speaking awkwardly due to meeting a celebrity.

Oppositional defiant disorder (ODD)

Oppositional defiant disorder (ODD) is a diagnosis sometimes given to a person, usually a child, who shows patterns of anger, irritability, and stubbornness. It's important to recognize that ODD is diagnosed more often in Black children, which suggests racial bias in how this diagnosis is made.

Social communication disorder (SCD)

Social communication disorder (SCD) is a term sometimes given to a person with social communication challenges, but where other autistic traits aren't seen.

Sensory processing disorder (SPD)

Sensory processing disorder (SPD) is diagnosed when a person is viewed as having sensory processing differences, but not other autistic traits.

Intellectual disability

An intellectual disability means a person's brain processes and understands information differently, often making tasks more challenging. They may need more time to learn and develop skills due to their unique cognitive timeline (see Chapter 9 for details). Autism is not an intellectual disability, but around 30 percent of autistic individuals also have an intellectual disability. Recognizing and supporting the strengths and needs of those who are both autistic and intellectually disabled is crucial.

Epilepsy

Epilepsy involves sudden "electrical storms" in the brain that lead to seizures. These seizures can trigger shaking, unusual movements, sensations, or brief loss of consciousness. While some with epilepsy have infrequent seizures, others experience them more often. Among autistic individuals, epilepsy is more common than in the general population, with about 20 to 30 percent of autistic people potentially facing epilepsy or seizures at some stage in their lives.

Sleep differences

Research suggests that there are more night owls among autistic people than the general population. These are folks who find their peak energy and concentration levels kicking in when others are winding down for the day. For them, sticking to a traditional sleep schedule may mean they don't get enough sleep, which can make it harder for them during the day.

REMEMBER

Sleep differences can also occur for other reasons. The connection between autism and sleep is complex, and more research is needed.

Myths and Misconceptions

In this section, we clear up some key misunderstandings about autism so that everyone can better support the autistic people in their lives.

Autistic people lack empathy

The myth that autistic people lack empathy is not only untrue but also harmful. Studies show that autistic individuals often feel empathy very deeply, but that it shows up in ways that others may not be familiar with — such as problem-solving or really good listening, which are important in emotional connections. To be inclusive, both neurotypical and autistic people should try to understand how each other expresses care and support.

Autistic people have superpowers

Saying "autistic people have superpowers" means well, but it creates false expectations and stereotypes and makes it seem that autistic people are less normal than anyone else. Everyone had challenges, and everyone has strengths — autistic people included. Some autistic strengths are quite amazing, but non-autistic people have quite amazing strengths too.

We're all a little autistic

The phrase "we're all a little autistic" means well, but it can make autistic people feel even more alone and misunderstood. It's true that anyone may have traits that are also common in autism. However, autism is a unique way of experiencing the world, with its own challenges and strengths, and autistic people want to be seen and understood.

Nonspeaking autistics are "not there"

The belief that nonspeaking autistic individuals lack intelligence or aren't aware of their surroundings is false. The full range of intellect and ability that exists in the general population exists in the nonspeaking autistic population as well. Just like you, nonspeaking (and minimally speaking) autistic people are aware of what's happening around them, including what others say.

The myth of severe (profound) autism

Using labels such as *severe* or *profound* autism stigmatizes autistic people, limits opportunities, lowers expectations, and can prevent them from getting needed support. Focus instead on understanding, accepting, and supporting each individual and their unique mix of challenges and strengths.

Parents are responsible for autism

The idea that parents cause their child's autism is not true and is hurtful. Research shows that autism is not caused by how parents raise their kids. This myth unfairly makes parents feel guilty and may delay parents in getting the help and support they need. Let's support families with understanding instead.

Chapter **7**

Understanding ADHD

I magine a young woman named Carrie. She's vibrant and has dreams as big as her energy. She's always known that she's a bit different; she's struggled with focus and friends sometimes call her "fidgety." One day, her doctor informs Carrie that she has ADHD. Things start to make sense, but before Carrie can fully understand what this means, her doctor paints a bleak picture: Carrie is "deficient." She's a problem to be fixed.

Carrie's self-image is shattered. She's no longer the creative, quick-thinking woman she thought herself to be. Instead, she feels marked by a glaring flaw, a misfit in a world that values how other people think. Carrie's confidence crumbles and is replaced by self-doubt. Anxiety creeps in, dimming the once bright spark in her eyes.

Carrie's story may sound dramatic, but it's far from fiction. It unfolds daily for many people diagnosed with attention-deficit/hyperactivity disorder, or ADHD. Now imagine if Carrie had been told about the unique advantages of her ADHD — her potential for out-of-the-box thinking, her bursts of hyper-focus, her innate creativity. Giving Carrie a modern and strengths-based framing of ADHD may have fueled Carrie's self-esteem, established her self-acceptance, and provided her the information necessary to manage her challenges as well as her strengths.

In this chapter, we discuss ADHD through the framework Carrie should have received. (Don't worry; she's okay. Carrie is just a character we created to show what it's like for many people with ADHD.) We cover what ADHD is, how to deal with its tough parts, how to make the most of its strengths, why we need people with ADHD, and how we can all support them.

Talking about ADHD

In the past, society often saw ADHD as a problem to be fixed, like telling someone who marches to their own beat that they're out of step. This old view focused almost exclusively on the challenges, such as getting easily distracted or being extra energetic, and often overlooked the cool aspects of ADHD, such as creativity and fast thinking.

Whether an ADHD trait is felt as a challenge or strength often depends on circumstance, understanding, and support. How people see these traits matters too. Khush from our author team shares her story: "As a clinician working with neurodivergent children, my ADHD traits were often praised in the workplace, but I was never evaluated based on them. Instead, I was mostly judged by the usual standards for how tasks should be done, which are based on neurotypical ways of working." Thankfully, Khush eventually found a workplace that supports neurodiversity, where she was encouraged to use her ADHD traits to improve the support she gives to her neurodivergent clients.

Shifting how we think about ADHD includes shifting how we talk about it — and that's important given that at least 7 percent of people may have an ADHD variation in how they think (for more stats on ADHD, check out Chapter 3).

Navigating outdated terms

The term *attention-deficit/hyperactivity disorder* feels a bit old-school from a neurodiversity point of view. The words within it such as *deficit* and *disorder* carry a certain cringy vibe. They're also not accurate, as they incorrectly make it sound as though something's wrong or missing from the ADHD person. It's like saying left-handedness or red hair are disorders just because they're not the majority.

"That's a terrible term. I see it more as a trait," says Dr. Edward Hallowell, a worldwide expert on ADHD in the 2018 YouTube video, "Dr. Edward Hallowell on Unwrapping the Gifts of the ADHD Brain." That doesn't mean that challenges are ignored. As Hallowell explains, "It can *become* a disorder. If you don't handle it properly it can become a disaster. But if you do handle it properly — 'unwrap the gift' as I like to say — then it can carry you to the heights."

Our modern understanding of ADHD is that it's another normal variation in brain wiring, not a flaw. However, the words we use to describe it haven't quite caught up. "ADHD" doesn't perfectly describe this brain variation, but it's the name we've got for now. Think of it like an old nickname that doesn't really fit anymore, but everyone still uses it. So, even if ADHD sounds outdated, we shouldn't get too hung up on the label. What's more important is shifting our understanding and conversations around ADHD.

Distinguishing between ADD versus ADHD

Before wading too deep into ADHD, it's important we clear up potential acronym confusion: ADHD, ADD, and AD/HD. They may look like alphabet soup, but they're just different ways to refer to aspects of ADHD.

>> **ADHD, mainly inattentive type:** This is what used to be called ADD. People with this type mostly have challenges around focus, organization, and task management. They aren't usually hyperactive.

>> **ADHD, mainly hyperactive-impulsive type:** People with this type are often very active and make quick decisions. They may also have some trouble focusing.

>> **ADHD, combined type:** This is the most common type and includes traits from both of the first two groups.

REMEMBER

Attention-deficit/hyperactive disorder (ADHD) is the umbrella term for several closely related experiences. While not all people with ADHD are hyperactive, we use ADHD in this book to keep things simple.

Understanding ADHD Traits

When it comes to ADHD, kids and adults can show it in different ways. Children with ADHD often have tons of energy, may wiggle a lot, or get distracted easily. They may do things impulsively or struggle to follow directions. As these kids grow into adults, these traits don't disappear, but they change a bit. Adults with ADHD may find it tough to plan things out, keep track of time, or stick with just one task. They may jump into conversations or feel fidgety. It's the same ADHD, just with a grown-up twist.

REMEMBER

The things that seem challenging for someone with ADHD can turn into strengths with the right help and understanding. A child's high energy? That can grow into great multitasking skills in adults. And if they seem restless, it may actually be because they have a really creative mind. When we recognize and support these qualities the right way, they can become strong points for both kids and adults.

How the ADHD brain thinks

ADHD brains are like speedy browsers with countless open tabs. They excel in problem-solving, spotting connections, and finding solutions others overlook by rapidly switching between these tabs. This quick thinking often makes ADHD folks spontaneous and fun in social settings. They don't just think outside the box; they may not even notice the box, as their minds are occupied with noticing various other amazing things.

A broad attention pattern

Focusing their attention? Well, for those with ADHD that's like trying to catch one tiny butterfly in a swarm of a thousand. This means that the ADHD person may miss details at times or may seem as though they're not listening. And while their energy and spontaneity are great for brainstorming sessions, they can struggle in structured settings or routines.

REMEMBER

Having ADHD doesn't mean someone is wrong or broken. The issue is that many of our systems, such as schools and workplaces, aren't designed for how someone with ADHD thinks. "It just kills me that there are standardized tests geared towards just one type of child," said actor Channing Tatum, who has both dyslexia and ADHD, in a 2013 *Vanity Fair* article. And, he's right. Whether it is our workplaces or our schools, we tend to build things without the ADHD brain in mind (for more on this, turn to Chapter 4).

But here's the thing: Even though society doesn't fully support people with ADHD, we desperately need their way of thinking and doing things.

"Many people with ADHD thrive in fast-paced, stimulating environments such as advertising, journalism, tech, entrepreneurship, and emergency first response services," said Margaux Joffe, the founder of the Kaleidoscope Society, an empowering community for women with ADHD, in a 2023 article on AdAge.com. "Neurodiversity is our most underutilized creative resource. Discarding stigma, embracing the brilliance of neurodivergent minds and addressing systemic barriers will open doors to a more imaginative future."

Flow state: ADHD and hyperfocus

It's a misnomer to think that the attention of those with ADHD is always broadly scattered. At times, it can be hyperfocused like a laser beam. Picture this: Someone with ADHD finds something that really grabs their interest. Suddenly, it's like they're in a tunnel — the world around them fades away, and they're zeroed in on this one thing. It's not dissimilar to the attention we give our partner when falling in love.

"When I'm focused, there is not one single thing, person, anything that can stand in the way of my doing something," said Olympian Michael Phelps in his book, *No Limits: The Will to Succeed* (Free Press). Phelps has openly discussed how ADHD has contributed to his success in swimming. Neurotypical people often call this the "flow state," and lots of money and time have been spent learning how to achieve it. Yet for many people with ADHD (as well as autism), this experience comes naturally.

In a 2014 article for *Today's Parent*, journalist Lisa Ling also shared an appreciation for this trait: "It has helped me. I can hyperfocus on things that I am excited and passionate about."

Being in a flow state is like finding the right groove where everything just clicks. You're hyperfocused and crushing it, whether you're working, creating art, or playing a game. It's awesome because you're productive, creative, and it just feels so good. But here's the catch: You can't be in this state all the time. Life's got other stuff — friends, family, eating, sleeping, other responsibilities — to pay attention to. If you're always in flow, you may miss out on those.

Khush on our author team explains her ADHD flow state this way: "Hyperfocusing on a project is incredible. It allows me to keep polishing something until it's outstanding. But I have to be careful or I might get too into it and lose track of time. At that point, I might miss a deadline or forget about other important stuff I need to do."

Executive functioning

Executive functioning is basically the brain's command center for managing tasks and controlling behaviors. You can think of it like the director of a movie overseeing such tasks as planning, organizing, remembering details, and managing time. And how we use our executive functioning is particular to each of us.

When you have ADHD, this director can be a bit unconventional. It's not that the director's absent; it's just that their style is more . . . improvisational. This can mean that someone with ADHD may struggle with organizing tasks, keeping track of time, or remembering all the details. But they may also hold a larger vision, preferring to jump right into the action without sticking to the script.

Even today, a lot of people get it wrong about how folks with ADHD manage tasks and make decisions. They mistakenly think that these people are just lazy or not trying hard enough. But on the social networking site X in 2022, Ross Pollard, a psychotherapist who has ADHD, points out something important: "The idea that people with ADHD are 'lazy' is laughable, considering how proactive and curious to learn many ADHDers actually are. But often the environment isn't supportive and there is no encouragement." This means that people with ADHD are often eager to do things, but don't always get the understanding and support they need.

Supporting ADHD thinking

TIP

The current world may not be designed for ADHD minds, but there's a lot that can still be done to meet them in the middle. Here are some ways to support ADHD thinking:

>> Create a routine but be flexible when needed.

>> Break big tasks into smaller parts so that they're less overwhelming.

>> Use timers and checklists to help keep track of tasks.

>> Make time for short breaks for movement to help focus.

>> Make work or study spaces quieter and less cluttered.

>> Focus on the strengths often seen in the ADHD mind, such as creativity and problem-solving.

>> And most important, be patient and try to understand how people with ADHD see and process things.

This approach can really help people with ADHD use their unique skills effectively.

How the ADHD brain uses energy

Are you familiar with the term *hyperactivity?* Think of it as having a supercharged battery. People with ADHD often have an abundance of physical energy that's ready to go. We used to think that hyperactivity was a bad thing. People with ADHD were often told to "sit still" or "stop moving." We'd dismiss them as "hyper" or "problem children" or "fidgety" or we'd say that they were "lazy" or "unmotivated" or simply had "ants in their pants." It turns out, none of these things were true.

Having a supercharged battery of energy is great for the ADHD person who has it. It helps them excel in tasks that require physical activity or a quick response. It can fuel creativity, personal drive, and the ability to tackle problems with enthu-siasm and vigor. This can often lead people with ADHD to achieve great things in various fields such as the arts, sports, and business.

Cynthia Gerdes, cofounder of the renowned Minneapolis eatery Hell's Kitchen, talked about her struggles with organizing tasks in the 2022 *ADDitude* magazine article, "Girl Power(Houses): Inspiring Women with ADHD. She said, "I couldn't cook. Even with a grocery list, I couldn't get the five ingredients that I needed." Despite this, her boundless energy helped make the restaurant a tremendous success, with local media praising Gerdes as a "whirling manifestation" of someone "unabashedly ADHD."

REMEMBER

If the supercharged battery of the ADHD person doesn't match your pace, then that's completely fine. They have a battery that suits them, and you have one that works for you. The key is to understand and appreciate how each other's batteries function, so everyone can achieve their goals successfully.

How the ADHD brain uses impulse

People with ADHD often experience a flood of ideas, feelings, and impulses all at once. For many ADHD brains, everything seems worth exploring. This can lead to some amazing creativity and thinking, but it can also make it challenging to focus on one task or to pause first to think things through.

Impulse control, or sometimes the lack thereof, is a big part of the ADHD experience. Think of it as having an overly enthusiastic dog inside your brain. You see something interesting, and whoosh, your brain-dog dashes toward it without a second thought like it's chasing a ball. Examples include blurting out an idea in a meeting before fully forming it, or suddenly deciding to reorganize your bookshelf at midnight.

How the ADHDer exercises impulse can be a great thing. It may help a person quickly think on their feet, dive into new projects with enthusiasm, and come up with creative solutions that others may not see. This ability can lead to amazing innovations, breakthroughs, and spur-of-the-moment brilliance. But it's also important for the ADHD person to learn how to effectively handle impulse as well. By learning to manage it, such as pausing to think or making lists and reminders, a person with ADHD can use their quick impulses more effectively. This approach lets them keep their creative spark, while also guiding their energy to turn quick thoughts into real, finished goals.

How the ADHD brain processes emotions

During the promotion of her movie *Barbie*, director Greta Gerwig revealed that she has ADHD — noting that her traits were easy to see in childhood, despite only receiving an official diagnosis as an adult. "I've always had a tremendous amount of enthusiasm. I was just interested in, like, everything. I had a really active imagination. I had a lot of really deep feelings. I was emotional," she told *The Guardian* in 2023.

Gerwig isn't alone. People with ADHD often feel emotions in a big, bold way. It's like having a hypersensitive emotional antenna. This means they can be incredibly passionate, experiencing joy and empathy on a grand scale. But this also means that tough emotions, such as frustration or sadness, can hit harder too. This intensity can sometimes lead to quick, impulsive choices, but think of it as living life in high definition. It's more vivid, more intense, and with the right support and understanding, it's completely manageable. It's a normal part of their vibrant journey through life.

REMEMBER

Someone with an ADHD brain can't just swap out the part of their brain that processes impulse and emotion. It's hardwired. The key here is to have the ADHD person understand and accept themself for who they are. Doing so helps them realize where impulse and intense emotion present a benefit, and where they may be a challenge in certain parts of their life. With this understanding, their brain can start to adapt and change. The impulses and deep emotions don't disappear, but the way the ADHD brain processes them becomes more flexible, offering benefits along the way.

How the ADHD brain communicates

The ADHD brain works in a unique and fascinating way when it comes to communication. It's kind of like having a mind that operates on multiple channels at once. People with ADHD often process and express thoughts in a way that's less linear and more associative. In practical terms, someone with ADHD may quickly switch topics in a conversation, drawing connections between ideas that don't seem related at first. They may also offer unique and creative insights, seeing solutions or angles that others may overlook due to their rapid and associative thinking.

From a neurodiversity standpoint, this style is just a different approach to processing and sharing information. It's not about being right or wrong; it's about variety. Imagine a conversation with someone who has ADHD as a journey through a landscape filled with various pathways and shortcuts. It can be unexpected and incredibly creative — and in the right settings, downright fun. This way of thinking and communicating can bring a lot of depth and originality to discussions.

How the ADHD brain socializes

How the ADHD brain handles socializing is pretty interesting. It's also totally normal, just in its own distinct way. When someone with ADHD is socializing, their brain is often running on all cylinders. This can mean they're super engaged, really energetic, and often quite creative when hanging out. They may be the one who brings up new topics, makes quick jokes, or offers a fresh perspective that others may not have considered — sort of like a DJ switching between tracks.

From a neurodiversity point of view, this is just a different social style, not a wrong one. It's great to have friends who can make get-togethers more lively. Sure, they may sometimes miss social cues or change the subject abruptly, but that's just part of their unique way of interacting. It's not about being impolite; it's how their brain naturally navigates social situations.

REMEMBER

Not every person with ADHD is super outgoing or the center of attention. Some may prefer to stay in the background while socializing. Even though their brains are making fast connections like a more extroverted person with ADHD, they may prefer to quietly watch and absorb what's going on.

Understanding ADHD sensory traits

Sensory processing differences can be a part of ADHD, though they're not a defining characteristic. Some people with ADHD may be more sensitive to sensory input such as noises, lights, or textures. This heightened sensitivity can lead to feelings of being overwhelmed or distracted in environments with a lot of sensory stimuli. Of course, when colors, music, or sensations are more vibrant, they can often provide a lot of joy (we go over sensory processing differences in more detail in Chapter 6).

Enhancing Daily Living

Living with ADHD can be an awesome experience that's full of creativity, energy, and unique perspectives. When people with ADHD get the right understanding and support, they can really thrive, turning their fast-paced minds into an engine for innovation and problem-solving.

But, let's be real: Navigating our modern world with ADHD can also be tough. Society often isn't set up for brains that work at warp speed or bounce around a lot. This mismatch can make school, work, or everyday tasks really difficult. It's a bit of a rollercoaster, with amazing highs and some tricky lows. But when ADHD is embraced and supported, it can lead to some pretty amazing adventures and achievements.

TIP

For people helping kids with ADHD, using visual schedules can be a big help. Show them how to use calendars, timers, and other tools to stay focused and organized. This can make it easier for them to take care of themselves and handle their daily tasks more smoothly.

Understanding trauma and ADHD

Living with ADHD in a world that's not really set up for ADHD brains can be pretty tough, and sometimes, it can even lead to trauma. As we talk about in this chapter, when you have ADHD, your brain works differently than most. But the world often expects everyone to fit into a certain mold, like sitting still for hours or paying attention to things that don't grab your interest. When people with ADHD can't be themselves or meet these expectations, they may face criticism, misunderstandings, or feel as though they're constantly failing. This kind of ongoing stress and feeling out of place can really take a toll, leaving deep emotional scars. The relentless effort to adapt or hide their true selves can be really exhausting and damaging for the person with ADHD.

TIP

For those with ADHD dealing with this kind of trauma, it's important to find healthy ways to cope. One big step is connecting with others who get it — whether that's support groups, friends, or online communities. Sharing experiences and realizing you're not alone can be a huge relief. Therapy can also be a big help, especially with therapists who understand ADHD and the unique challenges it brings.

Understanding RSD and ADHD

When someone with ADHD faces rejection or even thinks they might be rejected, their emotional response can be intense, like turning the volume way up on a song. It's not just feeling a bit bummed out; it's more like an overwhelming wave of hurt or sadness. This happens because their brain is wired to feel emotions super strongly, especially negative ones like rejection.

This is known as *rejection sensitive dysphoria (RSD)*. It's like having an emotional system that's tuned to pick up every negative signal, no matter how small. For someone with ADHD, a minor criticism or a joking remark can feel like a huge deal. While specific percentages can vary, it's not uncommon to see estimates suggesting that around 50 to 70 percent of people with ADHD may experience symptoms of RSD to some degree.

We love how René Brooks, creator of the *Black Girl, Lost Keys* podcast links understanding RSD and ADHD with successful dating. In the episode, "How to Make Rejection Sensitive Dysphoria and Dating Easier," she says, "When you are sensitive to rejection, but also impulsive, that can sometimes equal a recipe for disaster. ADHD is screaming at us to go all in, take the risk, and 'live a little.' Rejection sensitive dysphoria is telling us that we are going to get hurt really bad. Neither one of these is the correct approach."

What she's getting at is how the impulsiveness of ADHD and the fear of rejection from RSD can make dating tricky, and finding a balance is key. Brooks continues with this great advice: "That doesn't mean you have to take the spontaneity and joy out of dating, I promise. It just means that you shouldn't go to Vegas drunk and get married, and you shouldn't assume that they are going to leave you heart-broken and unable to go on."

RSD can also show up in physical ways as well, such as a fast heartbeat, sweating, an upset stomach, or feeling tense. This can really hit someone's self-esteem and confidence, making them doubt their own abilities and worth. Although unhealthy, a person may avoid situations where they could be rejected or fail. This may lead to pulling back from social activities, not wanting to try new things, or putting things off.

REMEMBER

RSD isn't about being overly sensitive; it's a real part of how an ADHD brain pro-cesses emotions. Recognizing RSD in ADHD helps in understanding why certain situations can feel so tough and why support and empathy are key.

Understanding masking and ADHD

Masking is like putting on a costume, but instead of dressing up for Halloween, you're dressing up your personality. For someone with ADHD, this may mean pretending to be highly organized and calm when they're actually feeling scat-tered and full of energy inside.

Why would someone with ADHD do this? Well, often it's because they feel as though their real selves won't fit in or be accepted. They may be super energetic, impulsive, or have a hard time focusing on just one thing, and they worry that others may not get it or may judge them. So, they mask these traits. They may try really hard to sit still, not blurt out what they're thinking, or force themselves to pay attention even when it's hard.

But masking can be exhausting. It's like trying to hold a beach ball under water all day. There you are, bobbing up and down, and you've got to put a whole lot of energy into keeping that ball beneath the waterline in order that no one else can see it. That's incredibly tiring. It takes a lot of effort to keep something down.

REMEMBER

If you have ADHD and you're tired of masking, there are healthier ways to cope. First, embrace your unique brain! ADHD brings creativity and energy that's totally worth celebrating. Find environments and people who get you and make you feel comfortable being yourself.

When it comes to giving up masking, cultural icon and ADHD advocate Paris Hilton offered this advice in her book, *Confessions of an Heiress* (Fireside): "The only rule is don't be boring and dress cute wherever you go. Life is too short to blend in."

TIP

Part of unmasking is getting rid of guilt. You shouldn't feel shame about using tools that help you stay organized and focused. To-do lists, timers, or apps can be lifesavers. Taking care of yourself is also key. Regular exercise, good sleep, and healthy eating can all help. Most important, be kind to yourself. It's okay to be different. Just keep being your awesome self!

Understanding burnout and ADHD

The phenomenon of ADHD burnout may be compared to "hitting the wall" while running a marathon (and one you didn't even know you were in). It happens when folks with ADHD have pushed themselves extremely hard for a long time in order to keep up with modern life's demands. Imagine trying to juggle a million things at once — work, school, social life, daily chores — and your brain is constantly on overdrive to handle it all. Over time, this can really wear you out.

The ADHD person may start to feel exhausted, both mentally and physically, and things they'd normally be able to do easily can suddenly feel like climbing a mountain. It's like the brain's batteries are drained, and even the simplest tasks seem huge. This burnout is the ADHD brain's way of saying, "Hey, I need a break!" It's important for a person with ADHD to listen to that signal and take some time to take care of themself.

TIP

To cope with ADHD burnout, set boundaries to avoid taking on too much, and break tasks into smaller steps to make them less overwhelming. Remember, dealing with burnout is less about pushing through and more about stepping back to recharge and regroup.

Nurturing one's wellness

The ADHD brain is always juggling a lot of thoughts, ideas, and emotions. So, taking care of oneself isn't just nice; it's essential. This means finding ways to relax, such as going for a walk, listening to some favorite music, and getting enough sleep.

"Exercise helps me meditate and get rid of my anxiety," said singer and Spice Girls member Mel B, who is multiply neurodivergent (dyslexia, dyspraxia, and ADHD), on *Paul McKenna's Positivity Podcast* in 2019. "It helps me focus on me for that one hour."

As with Mel B, many people with ADHD find meditation helpful because it offers a quiet place to focus, even when their minds are very active. This doesn't change their ADHD traits; it just provides new ways for them to process information and experience life.

REMEMBER

Everyone's mind works differently, and that's true for how people with ADHD may take to meditation. Sue Hutton, who leads neurodiversity-affirming mindfulness classes, put it this way in a 2020 article on Mindful.org: "We all have different brains and our own ways of responding to mindfulness practices." If the first method tried doesn't click, Hutton encourages exploring others. It may not be that you're "doing it wrong" she said. "It may be just how you're wired."

Maintaining one's wellness is also about resting, eating well, and maybe even finding a hobby that helps you unwind. And nurturing one's wellness can mean leaning on friends, family, or support groups for encouragement. If needed, therapy can be a great way to learn strategies to manage ADHD challenges and learn how to be confident in one's ADHD strengths.

Advocating for oneself

Self-advocacy means understanding who you are, making your own decisions, speaking up for yourself, and asking for what you need to thrive. It's a resource that anyone with ADHD needs.

In a 2019 article in *Elle UK*, actor Emma Watson, who has ADHD, put it this way: "I don't want other people to decide who I am. I want to decide that for myself." For folks with ADHD, that means speaking up and sharing needs with others, such as teachers, bosses, or friends. It also means confidently standing in your strengths.

REMEMBER

When you advocate for yourself, you're saying, "Hey, here's how my brain works, and here's what can help me succeed." It's not about making excuses; it's about getting the support you need. Similar to needing a ramp to access a building with a wheelchair, speaking up about your ADHD is about leveling the playing field and making sure you have a fair shot at success. So, don't be afraid to raise your voice and let people know what you need to thrive. Self-advocacy is a powerful tool for living your best life with ADHD.

Deciding on medication

When it comes to ADHD, medication is a bit like a tool in a toolbox — it's helpful for some, but it's not the only option. From a neurodiversity perspective, we understand that ADHD brains are wired differently, not wrongly. For many, medication helps to tune the brain's focus and energy levels, sort of like adjusting the settings on a camera to get the clearest picture. It can make a big difference in making daily life a bit smoother.

"Having ADHD, and taking medicine for it is nothing to be ashamed of; nothing that I'm afraid to let people know," said Olympic gymnast Simone Biles in 2016 on the social networking site X. She's right. We shouldn't be ashamed of taking medication, nor shame others for taking it. If this approach is helpful to a person, great!

REMEMBER

Medication isn't a one-size-fits-all solution. Just as some people prefer glasses and others go for contacts, the decision to use medication for ADHD is deeply personal. For some, other strategies such as therapy, lifestyle changes, or organizational techniques may be the primary way to go. It's all about finding the right mix that works for your unique brain. And that's okay — the goal is to support each individual in a way that celebrates their neurodiversity and helps them thrive.

Empowering Those with ADHD

When it comes to empowering folks with ADHD, it's all about embracing their unique way of thinking and giving them understanding and providing them with the right kind of support.

Shifting perspective

Toss out any old ideas that ADHD is a flaw — it's not! People with ADHD have brains that are wired to be creative, energetic, and full of out-of-the-box ideas. To really help them shine, we need environments that appreciate these qualities, whether it's in school, work, or at home. This means providing flexible workspaces, allowing for movement, and offering tasks that ignite their passion. Also, clear and concise communication helps a ton. And hey, don't forget about the power of positive reinforcement — a little encouragement can go a long way.

Taking ADHD seriously

You know how sometimes people joke about ADHD or toss around such lines as, "We're all a little ADHD sometimes"? This seems harmless, but it actually downplays the real challenges faced by those living with ADHD. It's not just about sometimes being distracted or hyper; ADHD is a genuine neurological variation that deeply affects people's lives. When we treat it as just a quirky joke, we miss the whole picture and make it tougher for those with ADHD to get the understanding and support they truly need.

Now, think about how you'd feel if a big part of who you are was dismissed as fake or not real. That's the sting people with ADHD often face. It's more than just feeling misunderstood; it can stop them from seeking help or accessing resources that can genuinely help. What we need is a shift in approach. If we start by believing and trying to understand each other's experiences, just like we'd want for ourselves, we're on the right track. It's about respect, listening, and being open to learn about experiences that may be different from our own. That's how we can truly support and respect those with ADHD.

Trusting people with ADHD

To support someone with ADHD, trusting them is key. People with ADHD often face doubt, judgment, or misunderstanding from others, even after sharing their diagnosis. It's vital to believe them when they discuss their challenges and strengths, as they know themselves best.

"I knew I had strengths that other people didn't have, and my parents reminded me of them when my teachers didn't see them," said David Neeleman, founder of JetBlue, in a 2005 article in *ADDitude* magazine. "I can distill complicated facts and come up with simple solutions. I can look out on an industry with all kinds of problems and say, 'How can I do this better?' My ADHD brain naturally searches for better ways of doing things."

REMEMBER

As Neeleman noted, trusting a person with ADHD is about acknowledging their strengths. However, believing in the challenges a person with ADHD faces is also a way of showing trust in their experience. For instance, understanding that they may have trouble with time management or staying organized helps build trust. This isn't about blaming everything on a person's ADHD or excusing them from responsibility. It's about understanding the other person so that you can provide better support.

Myths and Misconceptions

There are a good number of myths and misconceptions about ADHD. We tackle a few of the most common ones here.

ADHD people are unmotivated

The idea that people with ADHD are lazy or unmotivated is just a myth. In reality, people with ADHD often work really hard, but our modern world isn't adapted to how they use their attention, making it harder for them to stay focused or complete tasks in the same way others do. This isn't about being lazy; it's about the brain working differently. Understanding and offering support, rather than judgment, can make a big difference.

ADHD people are "scatterbrained"

The myth that people with ADHD are disorganized when they seem scattered or speak in unclear sentences is misleading. Often, when someone with ADHD is passionate about a topic, their intense enthusiasm leads to quickly jumping between thoughts. This isn't disorganization, but rather a manifestation of their deep interest and excitement. Their speech may be bursty or jumbled, but it reflects the whirlwind of ideas they have.

ADHD people are rude

People with ADHD are often mistakenly seen as rude due to their unique communication style. They may ask direct questions or interrupt, not out of disrespect, but because they process information differently and think quickly. Understanding and respecting these communication differences can improve interactions with those who have ADHD.

ADHD people share "too much" info

People with ADHD may seem as though they're oversharing, but this is because they usually want to give context or thoroughly explain their thoughts. They include details for clarity and expression, which may seem excessive to those unfamiliar with this style. This isn't about lacking boundaries; it's a distinct way of ensuring clear communication.

Chapter 8

Understanding Dyslexia, Dyscalculia, Dysgraphia, and Dyspraxia

You're standing in front of a classroom, hands on the podium, ready for your big presentation. You've prepared for this, but when you try to jot down some last-minute thoughts, your handwriting becomes hard to read, even to you. Your palms start to sweat. Fast-forward to dinner at a restaurant with friends. They ask you to read a funny text out loud, but you struggle through it as the room falls silent. You feel embarrassed. Then the bill comes, and as your friends talk about who owes what, a wave of panic washes over you.

These struggles are real life for people with dyslexia, dyscalculia, dysgraphia, and dyspraxia — neurodivergent conditions that affect how people think and learn and move. If you or your child find reading, writing, or math tough, you're not alone. With up to 20 percent of people potentially having one of these conditions, this chapter may be an eye-opener for you, whether you're learning about someone you know or discovering something about yourself.

Understanding Dyslexia

Dyslexia isn't a flaw; it's a variation in how the human brain processes information. Experts estimate that around 10 to 20 percent of people are dyslexic. That's a lot of us.

Though reading and writing may be challenging to those with this neurotype, dyslexia has nothing to do with how smart someone is. Some of humanity's best thinkers are dyslexic, and many people with dyslexia demonstrate higher-than-average skills in creative thinking and problem-solving (see Chapter 3 for some great examples).

"Dyslexia is just a different way of seeing the world, a different way of processing information, and a different way of coming up with great ideas," writes *Virgin Group* founder Richard Branson, who himself is dyslexic, in his blog post titled "This Is Dyslexia." "It's no wonder that the world is waking up to the amazing strengths that dyslexia gives you."

REMEMBER

Dyslexia comes with challenges, but it doesn't mean a person is broken or needs fixing. Recognizing this allows for the creation of effective strategies and supports to improve their daily life.

Dyslexia and human history

Dyslexia is often linked to reading and writing, but guess what? For most of human history, nobody could read or write. That's right, dyslexic people were just like everyone else when it came to those skills. So what were dyslexic brains up to before books? Thinking about this question can help us better understand and support dyslexic folks today.

Turns out, people with dyslexia likely had crucial roles in early human groups. They see the world differently, literally. Many dyslexic people are great at noticing details and patterns, and they pick up on things without needing words. Imagine how handy that would be for tracking animals or understanding unspoken social cues. These differences in thinking were probably invaluable for following animal tracks, recognizing patterns in nature, developing agriculture, constructing human settlements, and understanding what others meant without words.

In a 2010 article on Slate.com, Professor Alison Gopnik of the University of California, Berkeley puts it this way: "When nobody read, dyslexia wasn't a problem. When most people had to hunt, a minor genetic variation in your ability to focus attention was hardly a problem and may even have been an advantage."

Spotting patterns and understanding body language are still useful skills today, just like other strengths dyslexic people have. They were important in early human communities, and guess what? They still matter now. Dyslexic people have always had a key role in how humans get along and succeed.

Dyslexia in our modern world

Writing changed the game for humans. It let us save ideas, share wisdom, and build on what we knew. And it wasn't just about communicating with the people around us; writing suddenly let us send our thoughts through time and space. People far into the future could read what we wrote, learn from it, and add their own knowledge to ours. When you do that generation after generation, it leads us to our modern world.

We're built to work together and help each other. That's why scientists say we're a cooperative species. Instead of trying to do everything by ourselves, we split our roles and responsibilities between us. Hunters, gatherers, farmers, teachers, ship builders, bus drivers, and everything else. Not everyone needs to know how to build a ship, or drive a bus, or teach a class. As a result, it's human nature to build systems and tools that are "good enough" for *most* people who use them, but not all.

Humans are a "good enough" species, and writing is one of our "good enough" tools. It's brought us great benefits, but it doesn't suit everyone. Around 20 percent of us find writing systems tricky because our brains operate differently.

"Don't feel so bad," Kinko's founder Paul Orfalea once told dyslexic students while discussing his own dyslexia, as reported by the *Ledger* in Lakeland, Florida. "It's the language that's screwed up, not you."

Here's the cool part: Our written language may be screwed up, but we've learned a lot over the years thanks to our ability to write things down. Now, we're beginning to use that accumulated knowledge to make reading and writing easier for everyone, including folks with dyslexia. We're tweaking fonts, adjusting spacing, and making other changes to make it better for all. (Want to know more? Check out Chapter 24.) This is all possible because we now get how dyslexic folks think.

Decoding information

Humans use symbols, like letters, numbers, and images, to stand for different things. For example, the word *tree* is not an actual tree, but it represents the idea of a tree. The number 2 is not two of something, but it tells us there is more than one. An image of a heart represents feelings like love or the concept of health, even though it's not a real heart. We all have to learn what these symbols mean. This process of learning and interpreting symbols is called *decoding,* and it's how we make sense of letters, numbers, and images.

We love the explanation given by journalist Johnny Harris (for more on his story, see the sidebar later in this chapter). He points out that our world is full of symbols that don't have any meaning on their own. You may understand a symbol like the letter *a* because your culture and language have taught you what it means. However, if you're unfamiliar with a symbol like क from the Devanagari script or Ж from Cyrillic, you will not be able to decode them. However, to someone who knows these scripts, they convey a specific sound, just as *a* does for English speakers.

For the dyslexic person, decoding can be a challenge as the letters and words used to convey meaning aren't always designed with their brain in mind. "Your whole school experience is one giant lesson in decoding books, and symbols, and images to give meaning and knowledge to your brain," says Harris in a 2019 YouTube video titled "I'm Dyslexic." "If you're dyslexic, this whole decoding thing is hard."

Decoding how words are written

For the dyslexic person, decoding letters and written words can be difficult. Reading can feel like trying to untangle a knot, and spelling isn't any easier. Letters and words appear to shift, making it hard to understand what's written. Take the sentence, "The quick brown fox jumps over the sleeping dog." At first, it might appear to the dyslexic person as:

eTh qciuk nrbow fxo jumps oevr eht seelping odg.

We say "at first" because most people with dyslexia can eventually understand sentences, especially with coping strategies and support, even though it may take them longer than others.

This is a simplified example, and the experience of reading a sentence varies widely among dyslexic individuals.

Dyslexic people often employ different strategies to understand written language. For example, a dyslexic person may focus on the overall meaning of a sentence to figure out a tricky word. They could also use pictures or the way the text is set up to help them understand what's being said.

Decoding the sound of words

Of course, spoken words are also symbols. If someone says the word *tree,* it's not an actual tree that comes out of their mouth. Instead, the word *tree* represents an idea.

Dyslexic brains sometimes perceive spoken words differently. Depending on the circumstance, this can either be a challenge or an advantage. On the challenging side, people with dyslexia can struggle to pick apart the individual sounds of spoken words, which can make spelling, reading new words, and even pronouncing some words difficult. However, this distinct way of decoding spoken words can also enable the dyslexic person to recognize connections and patterns in speech that others may overlook.

For instance, in a book club where folks are discussing a complicated novel, someone with dyslexia may have a hard time reading the book quickly or spelling names right in their notes. But the dyslexic person may also be the first in the group to pick up on a hidden theme or story pattern that everyone else missed.

Another example: During a team brainstorm, a person with dyslexia may quickly notice a link between two ideas that seem unrelated when spoken by coworkers. This can offer a fresh way to solve a problem or come up with something new.

Utilizing working memory

Working memory is like your brain's notepad, where it briefly holds and plays with information to help you do things like read and solve problems. Dyslexic people often use their working memory in ways different than most, like relying more on auditory or visual cues, to process and understand information. This can have its advantages, but when applied to written language it can present as a challenge.

When reading, a person with dyslexia may struggle to keep track of several things at once. Usually, our brains quickly figure out what each letter means, what each word means, and what the whole sentence means, all at the same time. For someone with dyslexia, doing all of this at once can be tough. This can make understanding what they're reading harder and also slow down how fast they read.

REMEMBER

It's important to remember that just because a dyslexic person may have difficulty with their working memory notepad when it comes to written words, it doesn't mean they don't have strong working memory skills in other aspects of life.

For example, Callie, who has dyslexia, may find it challenging to read a complex paragraph aloud in class. She may struggle with understanding each word while also trying to grasp the overall meaning of the paragraph. However, when the class shifts to group discussion, Callie excels. She offers insightful comments and comes up with creative solutions to questions raised by the text, demonstrating her strengths in thinking and problem-solving.

It's key to remember that people with dyslexia often shine in other aspects important to learning, such as problem-solving or creativity.

Areas of dyslexic strength

For a long time, the challenges of dyslexia were mistook as signs of laziness or a lack of intelligence. Unfortunately, this perspective left generations of dyslexic people without the understanding to support.

"They just assumed that you were just lazy, or stupid, and all these other things," said actor Whoopi Goldberg, recalling her experience growing up with dyslexia in a 2014 interview with author Quinn Bradlee. "But the thing that crushed me more than anything was that I didn't understand how they didn't see that I was smart. I just couldn't figure things the way they were doing it."

Goldberg was smart, growing up to become one of only a handful of people to ever win an EGOT (an Emmy, Grammy, Oscar, and Tony award). Yet as a kid, the challenges with dyslexia she experienced were overlooked, as were her unique strengths. Like other neurodivergent conditions, dyslexia isn't just about the difficulties; it also brings its own strengths. These differ from person to person, but there are some commonly seen strengths in people with dyslexia:

>> **Seeing the big picture:** Many people with dyslexia are great at understanding underlying themes and broader perspectives. This helps in solving problems and being creative.

>> **Great with shapes:** Dyslexic people often have a talent for understanding how shapes relate, which is useful in jobs like architecture, engineering, and design.

>> **Creativity:** A lot of people with dyslexia are really creative. They're good at thinking unconventional ways, boosting problem-solving.

>> **Spotting patterns:** Some people with dyslexia are especially good at noticing patterns, which is a skill that comes in handy in areas like math or computer science.

>> **People skills:** While it's not about processing information, it's worth noting that many people with dyslexia are good at understanding and interacting with others.

>> **Telling stories:** Many people with dyslexia are really good at speaking and telling stories or giving talks.

>> **Learning differently:** A lot of people with dyslexia find it easier to learn when they can see, touch, or do something, instead of just reading.

LEARNING TO THRIVE: THE STORY OF JOHNNY HARRIS

Johnny Harris and Iz Harris are award-winning journalists with a shared passion for storytelling. While their work has taken them to remote locations around the world, closer to home they resemble many typical neurodivergent families. Johnny himself has dyslexia, and the Harris' children include an autistic son.

Johnny Harris's dyslexia went undiagnosed in childhood, a common experience for many dyslexic kids. In a 2019 video essay titled "I'm Dyslexic" posted on YouTube, Harris says, "None of us knew that it was dyslexia. During this time, I just knew I wasn't good at reading and I wasn't quote 'very smart.'"

Harris was lucky to have a mom who read to him throughout high school, helping him get past the texts that tripped him up. But soon he was off on his own and entering college, which Harris describes as his "nightmare" due to the amount of reading that was required. Not one to give up, Harris came up with his own plan: He recorded lectures and listened to them over and over while jogging or walking to campus. These clever strategies helped him grasp what he needed to know without having to dive into dense reading assignments. Still, he struggled with written texts and papers, but that's where Harris' knack for storytelling really kicked in.

"My second creative solution was starting to lean into visuals as my way of presenting things," says Harris. "So, instead of writing a paper for a project, I would say 'Hey. Can I make a video about it?' or 'Can I make an infographic?'"

It's hard to say where Johnny Harris would be without dyslexia. But, as Harris' creative problem-solving in college aligned well with dyslexic strengths, we'd like to think dyslexia helped make him the amazing storyteller that he is. And while Harris' condition wasn't diagnosed until long after his school years, his story provides valuable lessons on how to help dyslexic kids today.

Johnny Harris shows us that kids with dyslexia are just as capable as any other student. The problem is, our schools haven't been giving them the support they need (for more on that, check out Chapter 4). So, let's change how we do things. Picture classrooms where students with dyslexia aren't just getting by, but are actually doing great. Imagine that teachers had the information and resources they needed to adapt their teaching methods for dyslexic minds. It would benefit not just the dyslexic students, but everyone. This isn't just talk — it's something we can really achieve. Let's do it.

Understanding Dyscalculia

Picture yourself looking at a math problem and feeling as though the numbers are playing a confusing game with you, as though they're dancing around. This is what many people with dyscalculia experience, making even simple math tasks tough. Dyscalculia is when someone's brain has a tough time grappling with numbers and math — it's like the brain's gears just don't mesh well with numerical information.

REMEMBER

An estimated 5 to 7 percent of people have dyscalculia, making it a common variation in the way humans think and interact with the world.

People with dyscalculia often find everyday tasks related to numbers challenging. Things that seem simple to most, such as counting a collection of items, understanding discounts at the store, or reading a map, can be difficult for them. They may also struggle with managing money and sticking to a budget. In school or at work, tackling math problems or quickly making sense of data can be tough.

For example, a person with dyscalculia may get nervous when it's time to split the bill after dining out with friends. Most people can quickly figure out their share and add a tip, but for someone with dyscalculia, this can be really hard and stress them out. They may use a phone calculator, quietly ask a friend for help, or even offer to pay more than they should because they're worried they've miscalculated.

"Imagine the anxiety I felt as a new teacher every time I have to write out a date or number on the classroom board," says Russ Heslabeck, who has dyscalculia and who teaches middle school English in Baltimore. "I love my students, but kids at that age are quick to point out any small mistake their teachers make."

However, Heslabeck points out that accepting one's neurodivergent traits can make a big difference. "I now share my diagnosis with my students to help them see that our school's neurodiversity includes their teachers as well. I no longer feel that same anxiety because I'm open and honest with them about how my brain works."

A lot of the stress folks with dyscalculia feel comes from people not really getting their condition. Whether it's at school or work, not understanding the challenges they face can cause misunderstandings and extra pressure. By getting to know more about dyscalculia and what comes with it, we can make number-related tasks way less stressful (for more on that, turn to Chapter 24). This isn't just good for those with dyscalculia, it's great for everyone.

We didn't understand it for a long time

The role of dyscalculia in human history is still a bit of a mystery. Some folks think it might be the flip side of skills that helped us out a lot in the past, such as being good with words or thinking outside the box. Throughout much of our history, being good at socializing and creative thinking may have been more important for survival than being good at math.

A lot more is known about dyscalculia now than in the past. However, it wasn't until recently that experts started studying it as its own condition, separate from other learning challenges. Before dyscalculia was understood as its own neurodivergent condition, those with it were often widely misunderstood. As with dyslexia, those with dyscalculia were routinely labeled as lazy or not smart, and their struggles were blamed on either bad teaching or the student not trying hard enough.

Even after experts started studying dyscalculia, many teachers and schools didn't think it was a real condition. This lack of understanding made things worse for those who had it, causing low self-confidence and more stress. In short, not understanding dyscalculia often led to using teaching methods that made things even harder. Thankfully, that's beginning to change.

Different expressions of dyscalculia

Dyscalculia affects people differently. Some may struggle with turning math words into symbols — it's like trying to translate a language they don't speak. Others may have a tough time seeing how numbers and objects fit together, which makes something like measuring a real pain. Then there are those who can look at numbers all day and still get mixed up, messing with their ability to solve math problems on paper.

Some folks with dyscalculia get really confused by graphs or charts. And many find it hard to get the hang of even basic math ideas, such as adding. What's interesting, though, is that some people can be pretty good at certain parts of math, but still stumble in other areas.

REMEMBER

In addition, lots of folks are starting to see that dyscalculia can show up alongside other conditions such as dyslexia, attention-deficit/hyperactivity disorder (ADHD), and autism. This mix can make it tough to pinpoint exactly what someone is dealing with. The good news is, getting the full picture of how these conditions overlap helps us really get to know the person and provide more tailored support.

Difficulty with numbers

For people with dyscalculia, understanding basic things about numbers can be hard. They may find it tough to get the idea of how much a number is worth, compare numbers, or even tell time. It can also be difficult for them to picture numbers in their mind, which affects how they handle everyday situations that involve numbers.

Doing math

For people with dyscalculia, basic math such as adding, subtracting, multiplying, and dividing can be really hard. Even understanding what math symbols mean can be confusing. This makes solving even simple math problems a struggle. They may also have a tough time following the steps in math problems. But it's key to remember that people with dyscalculia can be really good at other things and should get help finding different ways to tackle math tasks.

On our author team, John has dyscalculia and shares the following insight: "I have a knack for understanding complex data, but if you were to ask me to add five plus three, I would still need to use my fingers to count out those two numbers by hand."

Talking about math

People with dyscalculia often find it tough to talk about math. It can be tricky for them to use or understand math terms. Explaining math problems in words or solving word-based math problems is often difficult too. However, this doesn't mean they aren't smart. They may just need different ways to understand and communicate about math. Using pictures, visual aids, and storytelling to explain math can help.

For example, imagine a person with dyscalculia struggling in a math class when faced with a word problem such as "What is the sum of two dozen apples and three dozen oranges?" While their classmates quickly turn it into a math equation, the student with dyscalculia may pause at words like *sum* and *dozen*. But when the problem is shown with a picture of groups of apples and oranges, it suddenly makes sense.

TIP

Using storytelling can be a helpful way to explain math concepts to people with dyscalculia. Stories can make math more engaging and relatable, helping individuals grasp abstract concepts more easily. By weaving math into a narrative, it can become more understandable and less intimidating. This approach can also tap into the strengths of dyscalculic thinkers, such as creativity and problem-solving, making math a more enjoyable experience.

For example, imagine a teacher teaching multiplication to a student with dyscalculia. Instead of using abstract numbers, the teacher says, "You have three toy cars, and each car has four wheels. How many wheels do you have in total?" The student can easily visualize their toy cars and count the wheels. This storytelling approach helps the student grasp multiplication by connecting it to a real-world scenario, making math easier to understand.

Working memory

Working memory is like your brain's notepad. It's what you use to work on several problems at once. We each use our working memory differently; no two of us use our notepad in quite the same way. Some people may be better at remembering certain things in certain situations, while others may struggle more. Working memory isn't the same for all, and the challenges and strengths associated can be different depending on the situation.

Like anyone else, people with dyscalculia have areas where they have strong working memories. But when it comes to using working memory for math, it can be tough for them. People with dyscalculia often struggle to hold onto numbers and steps in their head while doing math. They may only be able to focus on one math concept at a time. This can be challenging because math problems often have multiple parts or steps to remember.

TIP

When it comes to helping individuals with dyscalculia manage their working memory for math, teachers can implement strategies like breaking down complex math problems into smaller, more manageable steps. Providing visual aids, such as charts or diagrams, can also help individuals with dyscalculia visualize the problem, reducing the cognitive load on their working memory. In addition, allowing extra time for problem-solving and encouraging the use of calculators when appropriate can help.

Individuals with dyscalculia can employ various strategies to make math tasks easier. At school, they can take notes during math lessons to write down important numbers and steps, reducing the need to remember everything in their head. They can also use digital tools and apps to assist with calculations and time management and alleviate the strain on their working memory.

Understanding shapes and spaces

Understanding shapes and spaces is essential in math, but for those with dyscalculia, it can be a bit tricky. Tasks like measuring lengths or making sense of graphs may pose challenges. To illustrate, picture having a map with various locations marked as dots. Now, imagine you need to calculate the distance between two specific points on that map. For someone with dyscalculia, this may be tough, as estimating the distance just by looking at the map may not come naturally to them because of how their brains process spatial information differently.

REMEMBER

Many people with dyscalculia are terrific when it comes to tasks involving shapes and spaces, highlighting the uniqueness of everyone's strengths.

TIP

Teachers can make math less daunting for individuals with dyscalculia by learning about the condition and its traits.

Strengths associated with dyscalculia

Dyscalculia isn't just about challenges; it's also about unique strengths. Numbers may not be their strong suit, but people with dyscalculia have other talents. They often excel in verbal communication and creative thinking. So, while math can be a stumbling block, it's just one aspect of the diverse set of skills they offer. Here are some strengths commonly associated with those who have dyscalculia:

>> **Shapes:** Some are great at understanding concepts related to geometry.

>> **Patterns:** Some quickly spot which things are similar or different.

>> **Real-world math:** When it comes to applying math to real word situations, some have a strong knack.

>> **Solving problems:** Many are able to think creatively to solve problems in unique ways. In addition, many with dyscalculia often discover their own clever and unique ways to process numbers and tackle math.

REMEMBER

Knowing these strengths helps us support people with dyscalculia in school, work, and daily life.

Understanding Dysgraphia

People with dysgraphia find it challenging to write or even type by hand. While it can make tasks such as taking notes or filling out forms more difficult, it's not a measure of someone's intelligence or creativity. In fact, many people with dysgraphia excel in areas such as storytelling, problem-solving, and other forms of expression. Dysgraphia is just a different way of interacting with the world, and various strategies and tools can help manage challenges.

Dysgraphia can't be fixed. It's a part of human neurodiversity that experts think is shared by 5 to 20 percent of people. Many with dysgraphia also have other neurodifferences such as dyslexia or ADHD. While we should help support challenges associated with dysgraphia, it's important to acknowledge the strengths that come with it.

Organizing thoughts

Sorting out your thoughts can be hard if you have dysgraphia. If you have dysgraphia, you may need extra help, and that's okay. Speech-to-text apps or talking into a recorder can make it easier to get your ideas out. So, even though dysgraphia can make some tasks tough in regular situations, it also opens the door to creative ways of organizing and sharing your thoughts.

Spelling and grammar

A person with dysgraphia may have difficulty spelling words correctly, even if they know how to spell them verbally. For example, they may consistently misspell common words like *because* or *receive,* even after repeated practice and instruction. This spelling challenge can be frustrating but helpful tools like spell-check or speech-to-text can make writing easier.

Handwriting

People with dysgraphia often find handwriting tough. Their writing may be messy, uneven, or hard to read. This can make schoolwork such as note-taking and writing assignments challenging, as well as everyday writing tasks. The way dysgraphia affects handwriting can be really frustrating, making simple things that involve writing by hand harder.

Fine motor skills

Fine motor skills involve tiny hand and finger movements for tasks such as picking up little things, writing, or buttoning a shirt. People with dysgraphia may struggle with these. They may grip pens in ways that seem odd to others, form letters uniquely, or space words differently than most.

Working memory

Much like dyslexia and dyscalculia, people with dysgraphia have their own unique working memory profiles. While they may excel in certain areas, using working memory for writing tasks can pose challenges. This can make it difficult for people to hold in their head the correct spelling of words or the sequence of letters needed while composing written text. Those with dysgraphia may find it hard to remember the rules of grammar.

Areas of dysgraphia strength

Although we can't diagnose historical figures, author Agatha Christie talked openly about her challenges with spelling and handwriting despite being an otherwise private person. "Writing and spelling were always terribly difficult for me," said Christie in her autobiography. "My letters were without originality. I was . . . an extraordinarily bad speller and have remained so."

Despite struggles with writing and spelling, Agatha Christie left an indelible mark on literature. She authored 66 detective novels, created the longest-running play, and remains the best-selling novelist and the most-translated author ever. Her story reminds us that we all have hurdles to overcome. People with dysgraphia may find writing difficult, yet it's important to remember their strengths. Here are some common ones:

- >> **Talking:** Many are great at speaking and clearly communicating.
- >> **Creative thinking:** Many find creative ways to solve problems.
- >> **Eye for detail:** Some may pay extra attention to small details.
- >> **Reasoning:** Those with dysgraphia often have strong reasoning skills.
- >> **Learning in different ways:** Many people with dysgraphia learn better when they see or do something, rather than just reading about it.
- >> **Being empathetic:** Experiencing challenges often makes people with dysgraphia more understanding toward others facing difficulties.

REMEMBER

Dysgraphia has its challenges, but it also comes with unique strengths. By addressing the difficulties and tapping into those strengths, we can help people with dysgraphia succeed.

Understanding Dyspraxia

Your brain is bossy. It's constantly shooting out signals to make your body do things. For example, when you're holding a cup of tea and bringing it to your mouth — sure, it's your hand holding the cup and your arm moving it, but it's really your brain telling them what to do.

Up to 10 percent of people have dyspraxia, where communication between the brain and body is tricky, making smooth movements hard. Tasks such as tying shoelaces or playing an instrument can be tough, but those with dyspraxia also bring unique strengths and perspectives. In addition, dyspraxia is common among

50 to 80 percent of autistic people, but it's not universal (and although numbers vary, many people with dyspraxia display traits of ADHD as well).

Motor coordination

Remember that cup of tea? When your brain tells your arm to raise it up to your mouth, that's called *motor coordination.* It's your brain and body working in sync to get a task done. When you have dyspraxia, your brain-to-body communication can feel a bit like a game of Telephone where players whisper a message to one another in a chain, with the message usually ending up distorted by the end.

It's not that the message gets lost; it just sometimes gets a little . . . remixed. So, tasks that require a lot of motor coordination, such as buttoning clothes or riding a bike, can become difficult. For some people with dyspraxia, this happens a lot. For others, it occasionally presents a problem. But hey, everyone's brain is a bit different.

TIP

If you're working on strengthening motor coordination, try these strategies:

>> **Be patient.** Give enough time to complete tasks without pressure.

>> **Break tasks into smaller steps.** For example, focus on putting on socks before shoes.

>> **Use pictures or show how to do things step by step.** This makes instructions clearer.

>> **Practice regularly.** Doing something repeatedly can improve ability and confidence.

>> **Keep areas neat.** Less clutter can make it easier to concentrate and move smoothly.

Spatial awareness

Spatial awareness can play a part in dyspraxia too. It's like your body's GPS and helps you know where your body is in relation to other objects and helps you move about the space around you. For example, it helps you catch a ball or navigate through a crowded room without bumping into people or things.

For those with dyspraxia, this internal GPS can sometimes feel a bit glitchy. Imagine trying to catch a ball, but your hands are just a tiny bit out of sync. Or picture yourself walking through a room and misjudging the distance between you and the furniture, perhaps landing you with a bumped knee or a spilled drink. Of course, this doesn't mean that those with dyspraxia can't learn to navigate a room or catch that ball. No, not at all. It just may take a bit more focus and practice.

Speech and language

For some folks with dyspraxia, speaking can be tough. It's not that they don't know what to say; their brain just isn't sending the right signals to the speech muscles. Imagine a conductor trying to lead an orchestra where every musician is playing their own song. That's kind of what it's like. It's frustrating, yes, but with the right help and maybe some speech therapy, people with dyspraxia can learn to make their words come out more easily.

Strengths

People with dyspraxia can be really talented, even though they have some hurdles to jump. They often stand out in creative stuff like art or music. They're also clever when it comes to figuring things out and coming up with innovative ways to tackle tasks. Plus, they tend to be really good at understanding how other people feel, probably because they've faced their own tough spots.

"Neurodivergent individuals with dyslexia, dyspraxia, and ADHD have been educated in a system that was ill designed for them to thrive," said Tumi Sotire, known on social media as The Black Dyspraxic, in a 2020 *Forbes* article. "Therefore, people with these learning differences will display admirable qualities such as problem-solving skills and determination."

Actor Daniel Radcliffe, best known for his role as Harry Potter, has dyspraxia and highlighted this in a *Wall Street Journal* blog post: "The fact that some things are more of a struggle will only make you more determined, harder working and more imaginative in the solutions you find to problems."

Hey, here's a fun fact: People with dyspraxia may take much longer to get the hang of stuff, but all that extra practice can turn them into pros! They may build super-strong muscle memory in things that rely on repetitive motion like tennis, surfing, manufacturing, or playing in a band.

Myths and Misconceptions

There are a lot of myths and misconceptions about the things we talk about in this chapter. We take a look at some of them here.

Dyslexics see words backward

The myth that folks with dyslexia read words backward is just that — a myth. Dyslexia is really about struggling with the sights and sounds of language, not flipping words around. If you have dyslexia, you may have a hard time reading because matching sounds to letters can be tricky. Getting what dyslexia actually is allows us to provide better support.

Dyslexia is only about reading

Dyslexia isn't just a reading issue; it's a unique way of processing language. Sure, it can make reading, spelling, and writing tricky, but it also brings strengths like creative thinking and problem-solving. So, it's more than just reading difficulties; it comes with its own perks and challenges.

Dyscalculia is math anxiety

Dyscalculia isn't just about being nervous over math. Sure, if you have dyscalculia, you may feel anxious about math stuff, but it's actually a specific learning challenge. It's about having a tough time understanding numbers, math concepts, and symbols. Math anxiety is different — it's that sweaty-palms feeling some get when dealing with math, and it can happen to anyone, even if they don't have dyscalculia.

Dysgraphia is poor handwriting

The idea that dysgraphia is only about messy handwriting is wrong. It's a condition that impacts spelling, expressing thoughts on paper, and fine motor skills, making writing difficult. Recognizing dysgraphia's broader effects is key to providing proper support.

Dyspraxia is just clumsiness

Dyspraxia is not simply clumsiness. It affects your movement coordination, making various tasks difficult. It's about physical actions like catching a ball or tying laces; but it also affects speech, organization, and planning. Dyspraxia is not a lack of effort or focus; it causes the brain to process movement differently.

You're not that smart

Having dyslexia, dyscalculia, dysgraphia, or dyspraxia doesn't mean you're not smart. Lots of smart people have these experiences, with many achieving amazing things (see Chapter 2 for more). Rather than dismissing someone's intelligence, focus on what you can do to empower them.

You're just lazy

People often wrongly think that the things we discuss in this chapter come from being lazy or not trying hard enough. Not so. These conditions are about how the brain works and affect things like reading, math, writing, and movement. People with these conditions usually work really hard on these things.

You'll grow out of it

It's a myth that the conditions we talk about in this chapter disappear in adulthood. These conditions continue throughout a person's life. Understanding that fact helps ensure people get the appropriate support they need at all stages of life.

Chapter **9**

Understanding Associated Conditions

I n this chapter, we journey into the intriguing world of neurological experiences that go beyond traditional categories of neurodivergent variations (such as autism, ADHD, or dyscalculia). Imagine conditions like Tourette's, bipolar, OCD, cerebral palsy, or multiple sclerosis — each one a unique chapter in the individual's life story, often misunderstood or emerging unexpectedly. Then there are lifelong companions like intellectual disability, or the plot twists nobody anticipates: a brain injury, a stroke, or the shifts in how you use your brain that may unexpectedly come with aging.

Accepting neurodiversity means supporting and celebrating people for their differences. It's about recognizing the value and potential in everyone, using their unique perspectives to enrich our communities, and creating inclusive environments for everyone to thrive. This chapter encourages a shift in perspective, acknowledging the incredible potential in every mind and exploring ways to build an inclusive society dedicated to helping every individual live their best life.

Understanding Tourette's

Tourette's (also referred to as *Tourette syndrome*) is a condition that's still a bit of a mystery — doctors and scientists don't fully understand what causes it. But what is known is that each person with Tourette's has their own unique brain that works a little differently. Of course, you don't have to know everything about a person's brain in order to support them. Adopting a neurodiversity perspective toward people with Tourette's means focusing on acceptance, recognizing their experiences, understanding the challenges they may face, and leveraging their strengths.

Breaking down tics

Imagine your body sometimes does things without checking in with you. These are called *tics*, and they're like unexpected guests popping up unannounced. Tics are sudden, brief, repetitive movements or sound. These can be simple and fleeting, like an occasional eye blink or throat clearing that comes and goes and doesn't persist for a long time.

REMEMBER

You don't have to have Tourette's to experience a tic. It's not uncommon for a person to have a tic at some point in their life, especially during childhood. However, people with Tourette's tend to experience tics pretty regularly.

Physical tics

When we're talking about Tourette's, we're dealing with two main types of tics — physical and vocal. Physical tics are movements that a person can't completely control. They can be simple, like blinking a lot or shrugging shoulders, or more complex, like a series of movements.

Pop singer Billie Eilish, who has Tourette's, shared with David Letterman on his Netflix series *My Next Guest Needs No Introduction* that her condition includes subtle things that wouldn't be obvious to a casual observer. "These are things you'd never notice, like, if you're just having a conversation with me."

Vocal tics

Vocal tics, on the other hand, are sounds that just pop out, unplanned. They can be as straightforward as throat-clearing or as complex as saying full words or phrases. It's like your voice has a mind of its own and sometimes it decides to throw out sounds or words without your permission. And despite what many think, it's pretty rare for these sounds to be swear words; that's just a common myth about Tourette's.

Both types of tics can come and go and change over time, and they're not something that folks with Tourette's do for attention or because they're feeling nervous. Tics are a result of their brains communicating with their bodies, sending out signals that turn into these unexpected actions or sounds.

Recognizing social difficulties

Tourette's can make social situations a little more complex. It's not that folks with Tourette's lack social skills — it's that an unexpected tic may interrupt a conversation or draw attention. Picture yourself in the middle of a story when a tic suddenly jumps in. That can feel awkward and, let's be real, people don't always get it; some might even be a bit rude. This can lead to feeling on edge about social interactions. But, when people get it and offer support, folks with Tourette's can handle these moments much more easily.

Realizing strengths

Folks with Tourette's bring a set of strengths to the table, shaped by what they've been through. They tend to be very strong because they've had to deal with difficult things. They're usually good at understanding how other people feel because they know what it's like to be misunderstood. They can also focus really well on tasks, sometimes better than others. And their way of seeing the world can lead to really creative and different ideas. Like anyone, they have their own set of skills and talents in addition to those forged by their disability experiences.

Understanding OCD

OCD stands for obsessive-compulsive disorder. You might've heard people say "I'm so OCD" when they're just being neat or particular, but OCD is not about being super tidy or liking your books organized by color. It's a real, often intense condition that involves two main things: thoughts (obsessions) and actions (compulsions).

The prevalence of OCD varies worldwide, but studies generally estimate that about 1.2 to 2.3 percent of the population will experience OCD at some point in their lives. In the United States, that means nearly 8 million people may live with OCD (roughly the population of the state of Arizona), which means that it's fairly common. And OCD can affect anyone regardless of age, gender, or cultural background.

Demystifying obsessions (thoughts)

We all have thoughts, but here's the deal: OCD is a condition where people experience recurring thoughts that can cause a lot of anxiety. These types of thoughts are called *obsessions*. They may be about germs, fear of making mistakes, uncertainty about whether you turned something off before leaving your home, or even unwanted thoughts that can be disturbing. These aren't just ordinary worries; they're intense and can really interfere with someone's life. For the person with OCD, they suddenly pop into their mind, often repeatedly, without them wanting them to.

Though these worries may be conscious thoughts, sometimes they can manifest more as a feeling. A person with OCD often understands that their obsession doesn't make logical sense, but it still feels very real as though something needs to be addressed. This can be stressful and confusing and feel as though their brain is telling them two different things.

On our author team, John experienced frequent OCD episodes as a child that mostly faded as he got older. However, rare episodes still occur, causing him significant frustration because he recognizes the lack of logic behind them. John explains, "My OCD mainly shows up when I rent a car. I'll check the driver's door six or seven times when walking away, even though I know it's locked. During times of high stress, I'll also repeatedly check if I've closed my bedroom door when leaving the house. I'm aware that I've secured these doors, but I still feel compelled to double-check."

Demystifying compulsions (actions)

The need to act on an obsession is the second part of OCD. These actions are called *compulsions*. Think about what you do when you have an itch. You scratch it, right? For someone with OCD, their obsessions are like itches that they feel they need to scratch in order to find relief.

Compulsions are different for everyone and depend on the situation. For example, someone may need to wash their hands seven times to feel clean or check the stove repeatedly to be sure that it's off. Another person may need to recount things to ensure their accuracy, or do something over and over again until they feel they've done it just right. You may ask, "But don't we all double-check things sometimes?" Sure, but with OCD, these actions go beyond the occasional double-check and become a ritual that feels impossible to control.

People with OCD often know that performing these acts won't actually do anything practical, but they still feel the need to do them. It's sort of like when your doctor tells you not to itch an insect bite or a sunburn. You *know* not to do it, but you feel you still have to scratch that itch. Just as resisting the urge to scratch can be hard, for someone with OCD, it's incredibly challenging to ignore compulsions.

Pop star Katy Perry has spoken candidly about her OCD, mentioning her worries about germs and her need to arrange things alphabetically. In a 2010 article in *Q* magazine, she shared a specific example of how it affects her: "My worst one is if I see a pair of sunglasses with fingerprints on them. Truly, I cannot stand it. I go ape. 'Oh my God, I can't breathe! Wipe them clean! Agghhh!'" This shows that even minor things, like fingerprints on sunglasses, can cause intense distress and create an overwhelming urge to take action for someone with OCD.

Conflict with daily life

OCD isn't about quirks or personality traits; it's a serious experience that can be incredibly challenging for those who live with it. In certain situations, the brain gets stuck on a loop of doubt and fear, and breaking that cycle can be really tough without help. While some with OCD find it touches many aspects of their life, others may be impacted in just one or two areas in which their brain gets "stuck" (like checking a front door to make sure it is locked, or making sure a light is turned off). No matter what the individual experience is, getting out of this cycle often requires assistance.

Footballer David Beckham shared his experience with OCD in a 2023 Netflix documentary: "I have to have everything in a straight line or everything has to be in pairs." His wife, fashion designer and Spice Girls member Victoria Beckham, finds a bright side in his nightly cleaning rituals, which leave their house very clean. However, Beckham himself admits that his compulsions to clean can be exhausting. "Everything has to be perfect," he says.

Applying empathy to social difficulties

Living with OCD can make social situations a bit of a tightrope walk. Imagine that you're trying to focus on a conversation, but your mind keeps nagging you about whether you locked the door or turned off the oven. It can be really distracting. This may lead to checking your phone for reassurances or even stepping out to handle a compulsion.

In addition, there's often an internal embarrassment about these compulsions, a fear that others will notice and judge. This feeling can add extra stress to social interactions, making the person with OCD feel more isolated. It's not that they lack interest in the people around them; it's just that their brain often diverts their attention elsewhere, and the fear of being seen as odd can complicate socializing unnecessarily.

Finding support

Treatment for OCD often works well. This may include therapy, medication, or a combination of both. And just like other experiences, it's crucial to find support that gets a person's specific needs. "Even if you go to a therapist or psychiatrist, they may not be an (OCD) expert specifically," said comedian and game show host Howie Mandel, who has OCD, in an interview with Today.com in 2023. However, with informed support, many people with OCD can manage their thoughts and compulsions in a way that provides them relief and allows them to live full, productive lives.

Valuing strengths

People with OCD often excel in detail and precision, making them great at jobs needing accuracy, such as research or editing. They can turn their need for routine and structure into being productive, and their drive to finish tasks shows resilience. Despite the challenges of OCD, these qualities highlight their potential beyond the condition.

In an episode of his YouTube series *The Tim Ferriss Show*, entrepreneur and author Tim Ferriss stated, "I can scan a document that four lawyers have looked at and find stuff," highlighting his exceptional attention to detail, a trait he associates with his OCD. However, Ferriss also underscores the importance of maintaining equilibrium and seeking adequate support for managing the condition.

TIP

To support someone with OCD, it's crucial to understand and respect their experiences without judgment. Offer a listening ear and encourage them to seek professional help if they're not already doing so. Creating a patient and accommodating environment, whether at home or work, can make a big difference. Simple actions such as acknowledging their feelings, being flexible with routines, and learning about OCD can foster a supportive atmosphere that empowers them.

Understanding Bipolar

Bipolar (also referred to as *bipolar disorder*) is a health condition that affects a person's mood, which can swing from really high energy and activity levels to very low, sad periods where they feel down or lose interest in daily activities.

The high mood is called *mania,* and the low mood is called *depression,* and the shifts between these moods can be intense but are manageable with the right support and treatment.

It's fairly common for individuals with bipolar to be unaware of their condition, often believing that their mood fluctuations are shared by others. "When I was diagnosed with bipolar disorder the year I turned 50, it was certainly a shock," said journalist Jane Pauley in her PBS special, *Take One Step: Caring for Depression with Jane Pauley.* Despite Pauley's longstanding presence on American television, neither she, her coworkers, nor her audience suspected she had the condition.

Doctors and scientists are still learning about why bipolar happens; it's likely a combination of genes, life experiences, and how the brain is built. What is known is that roughly 2.4 percent of people worldwide navigate life with bipolar. Even without all the details, you can still use a neurodiversity approach to support those with bipolar by embracing their unique traits, creating supportive spaces, and being adaptable to their changing energy and moods.

Understanding shifts in mood

Imagine you're on a boat on the ocean. For most people, the sea of emotions has its ups and downs, but it's relatively manageable — like gentle waves. With bipolar, however, the waves feel more like a storm for some (that's Bipolar I), and for others, they feel choppy water with only the occasional squall (Bipolar II).

Starting with Bipolar I: Imagine your mood is a line graph. For most people, that line stays relatively steady day-to-day. But if you have Bipolar I, your line graph has some serious peaks and valleys. The peaks are called *mania* — times when you may feel you have really high energy and can do anything, need less sleep, and make risky decisions. The valleys? Those are the depressive episodes — when you may feel very down, lose interest in things, and just feel sluggish and heavy.

Bipolar II is kind of like its cousin. You've got the same valleys, those periods of depression, but the peaks are different. Instead of full-on mania, you have what's called *hypomania.* Think of it as mania's less-intense sibling. You may feel more energized and productive, but it's not as extreme and doesn't usually mess with your ability to function.

REMEMBER

How do you show respect and empower those with bipolar? First, by recognizing that they're not defined by these highs and lows. They have talents, interests, and dreams like anyone else. Second, support can be vital. This may mean being patient and listening, or helping them find good treatment and sticking with it. It's also empowering to create environments — at work, at home, in society — that are understanding and flexible.

Understanding impulse

People with bipolar may sometimes act on impulse, especially during phases when they feel overly energetic. They may suddenly decide to do something risky or unusual for them without thinking it through. Not everyone with bipolar will do this, but it happens often enough to be seen as a notable part of the condition.

Finding success in health management

Managing health when you're living with bipolar involves finding the right balance of treatment, self-care, and support. It's about being proactive — keeping regular sleep patterns, staying on top of medication, and being tuned into your own triggers. Doing this can really pay off. For instance, Stephen Fry, a well-known actor and writer, has spoken about how finding the right treatment and sticking to it has been crucial for his well-being. Similarly, actor Catherine Zeta-Jones has also discussed how professional treatment and personal management strategies have helped her thrive.

Comedian Maria Bamford, who has both bipolar and OCD, said in a 2019 interview with bpHope.com, "Friends and family are hugely helpful, and I'll always call my psychiatrist if things get dicey." Bamford also mentioned incorporating music, exercise, and reading into her routine as key components of her wellness strategy.

REMEMBER

Thriving with bipolar may involve a combination of strategies such as sticking to a regular sleep schedule, maintaining a balanced diet, or taking prescribed medications. Regular exercise and mindfulness practices, such as meditation, can also play a crucial role. With consistent management, people with bipolar can lead robust and fulfilling lives, harnessing their periods of high energy while maintaining stability. It's not about erasing the highs and lows but about managing them in a way that allows for thriving, not just surviving.

Being aware of social difficulties

Living with bipolar can sometimes throw a wrench into social situations. Imagine your mood is a seesaw: When it's up, you may feel super outgoing, maybe even too much, which can overwhelm friends or make things unpredictable. Then, when the low moods hit, it may feel impossible to even start a conversation, or you may withdraw, making it hard to maintain those connections you value. It's a balance game, and finding people who get that your social energy can vary — without judgment — makes a huge difference.

TIP

If you have bipolar, being honest with friends about how your mood changes can affect hanging out can go a long way. It helps them understand and prepare. It's smart to have some downtime after social events or a trusted person who'll give you a nudge when you need a break. Good friends who listen and adjust without making a fuss are crucial. With the right crowd, dealing with the social side of bipolar disorder gets a lot easier.

Realizing strengths

People with bipolar experience a unique set of strengths that can include heightened creativity, empathy, and resilience. Many have discussed how their periods of high energy can translate into bursts of productivity and creative work. The capacity for deep feeling may also foster a strong sense of compassion and connection with others. While bipolar certainly comes with its challenges, many find that it can also contribute to profound personal growth and an enhanced ability to bounce back from life's ups and downs.

Understanding Cerebral Palsy

Cerebral palsy (CP) isn't a one-size-fits-all condition — it's a bit like a personal fingerprint for each person's brain. It's a neurological diversity that shows up because of how a brain develops before or after birth, affecting movement and muscle coordination. People with CP may move or communicate differently, but that's just a slice of their story.

REMEMBER

In a world that's learning to embrace neurodiversity, CP may be seen as a distinctive way the brain interacts with the body. By framing it this way, society can get better at building spaces, communities, and tools that aren't just inclusive by design but are actually enhancing everyone's ability to contribute and participate.

Sascha Bittner is a policy expert whose work has benefited the lives of millions of families across the United States. As a quadriplegic with a speech disability as a result of cerebral palsy, Bittner speaks in a slow and deliberate manner while chairing government bodies or discussing issues with policymakers and reporters in places such as the White House. To accommodate those who are unaccustomed to her pace of speaking, Bittner often uses an assistant to repeat what she says at a more rapid pace.

"I was able to attend and graduate UC Berkeley because I had someone getting me up each morning, dressing me, and taking me to the bathroom," said Bittner in her Medium article, "Medicaid Has Made My Life Possible" in which she discusses how accommodations have empowered her success. "I was also able to provide disability awareness programs to thousands of students in the Bay Area because of the attendant care that Medicaid paid for." However, we'd be wrong to think that these things only help Bittner in her life. Because these accommodations are essential in Bittner's advocacy and policy work, they ultimately benefit thousands of additional people every single day. You may be one of them without even realizing it.

REMEMBER

When we talk about applying a neurodiversity frame to cerebral palsy, it's not just about ramps, attendants, and accessible doors — it's about attitudes that recognize that a person with CP has strengths, perspectives, and talents that are valuable and worth celebrating. So, understanding CP in the context of neurodiversity nudges us all to think about how we can shift our environment to support everyone's unique strengths.

Understanding Multiple Sclerosis

Multiple sclerosis or MS, is a condition where the immune system, which normally protects us, gets a bit mixed up and starts to attack the body's own central nervous system. Think of the nervous system as the body's internal internet, sending signals from the brain to different parts. In MS, the protective coating around the nerve fibers is damaged by the immune system. It's like the insulation on wires getting stripped away, causing messages to slow down or not get through at all.

REMEMBER

MS is different for everyone. For some, it's a nuisance. For others, it can be quite disabling. People with MS may experience a range of symptoms such as fatigue, difficulty moving, or changes in sensation. Because the condition can be unpredictable, it's like each day is a new navigation map, where the routes and road conditions keep changing.

Actor Selma Blair describes part of her experience with MS on Instagram this way: "I fall sometimes. I drop things. My memory is foggy, and my left side is asking for directions from a broken GPS." Since disclosing her MS in 2018, Blair has become a disability advocate and has empowered numerous people to seek diagnosis and understand their own MS.

"I was sitting in Selma's living room, our children playing, and I told Selma I'd been having this weird tingling in my feet," said actor Christina Applegate, who now proudly displays her walking cane on various red carpets, in a 2023 article about Selma Blair for British Vogue. "She said, 'You must get tested for MS.' [Even my doctor doubted it] but there it was. In essence, because of her, I'm going to have a better quality of life."

In their responses to their own MS, Blair and Applegate demonstrate the value of applying a neurodiversity frame to muscular sclerosis. It's about recognizing that just as humanity has a vast range of cultures, personalities, and talents, it also has a wide spectrum of neurological experiences. Sure, there are challenges, but there's also resilience, insight, and perspective that come from navigating a world that's not always set up for those with MS.

REMEMBER

Understanding and support go a long way. Flexibility in workplaces, inclusivity in communities, and advances in medical treatments aren't just good for those with MS — they make society richer and more robust for everyone. That's something worth caring about.

Understanding Intellectual Disabilities

Think of the human mind as a vast library with an endless array of books and information. Now, for some, the librarians in this library are really fast. For those with an intellectual disability (ID), the librarians may move at a more leisurely pace — and that's okay. The books are still there; it may just take a bit longer to find and read them. The mind with ID isn't chaotic or confused; it's simply organized in a way that works on a different schedule.

REMEMBER

People with ID may learn and develop skills more slowly than others. But here's the crucial bit: They can and do learn, grow, and actively participate in life. They're as "normal" as anyone else because, really, "normal" is just a setting on a washing machine, not a measure of a person's worth or abilities (we borrow that phrase from Whoopi Goldberg, who you can read more about in Chapter 8).

The neurodiversity frame is about embracing all the different ways brains can work. It's recognizing that someone with ID isn't broken; they're a variation in the vast spectrum of human cognition. Just as some of us are great at math and terrible at cooking, or amazing visual artists but can't sing a note, people with ID have their own sets of skills and challenges.

WARNING

Many people with intellectual disabilities need help with their daily routines. But it's important to understand that needing support doesn't mean they're dependent. It's just a part of how they live their lives.

Support is *not* about deciding things for someone; it's about empowering them to do what they want to do. It's giving them the tools to build their own best life. Everyone needs help now and then, from a web search to a friend's advice. For those with intellectual disabilities, the support they need may be different, but the goal is the same: a fulfilling life.

Miguel Tomasin is the band leader and drummer of the popular Argentinian underground band Reynols. Tomasin, who has Down syndrome and intellectual disability, receives frequent support from his mother in his work. In a 2019 interview with *TheWire* magazine, Tomasin said, "I don't go anywhere without my mom." This partnership has allowed him to release over 100 albums throughout his career. "Playing with [Reynols] is great. There are lots of fans. I sign autographs everywhere, in the taxi, in the market," he said.

REMEMBER

Respecting people means including everyone as a regular part of society, no matter how much support they need. It's about valuing and helping everyone find their own kind of success. Embracing neurodiversity is about celebrating what makes each person different and recognizing that as a key part of what makes us all human. Everyone has something special to offer.

WHO DOESN'T LOVE A BURRITO?

Whenever John from our author team travels to New York City, he follows a tradition. "When I visit my friends Amitesh and Cody in Brooklyn, we always get Taco Bell. It doesn't matter that their neighborhood is filled with world-class restaurants and celebrity chefs. It just makes us endlessly happy to sit, talk, and catch up over a shared meal of fast-food burritos," John says.

If you want a burrito, you should be able to get one. This simple idea is also about respecting capacity of people with intellectual disabilities for self-determination. You see, many people with intellectual disabilities live in supportive settings (often called group homes or care facilities). These places are meant to offer support, help with daily needs, and offer a sense of community. But sometimes they don't live up to that goal. Instead of helping residents make their own choices, they treat them as though they can't decide for themselves. That's not how it should be.

So, how do you know whether a group home is really letting its residents live like anyone else? People in the intellectual disability community suggest using The Burrito Test. It's a simple test to pass: Can someone living there heat up and eat a microwave burrito at midnight if they want to? When seeing this question pop up as a comment on his personal blog, Dave Hingsburger, the founder of the *International Journal for Direct Support Professionals*, decided to explore it further. "This comment has resulted in me having several conversations that I would never have had before. Most people I spoke to, from several different agencies, after careful consideration said the answer would be, in most cases, 'No'."

Everyone has the right to make their own choices and be in charge of their own lives. This includes adults who need extra help. Hingsburger said in his article, "Burritos and Cherry Pies" in *The Direct Support Workers Newsletter* that "our job isn't to be a substitute parent whose role is to forbid the people we support from things we deem bad for them or inconvenient for us."

This idea is important when you think about the help and services you get for yourself or others. It's like asking whether you're free to grab a microwave burrito when you feel like it. Can you make your own choices? Everyone should be able to say Yes to these questions, no matter how much support they need day-to-day.

Aging and Acquired Impairments

As we age or face challenges such as brain injuries or strokes, our brain's functioning can shift. These changes are a normal part of life, and many individuals go through them. For instance, following a stroke, someone may need to relearn skills such as speaking or walking. And as we grow older, our memory may work differently than before — often our memories become sharper in some respects (the past) and worse in others (current events). While these changes can feel intimidating, they're actually quite common.

Many of us worry about the effects of aging or the aftermath of brain injuries or strokes. We fear becoming overly reliant on others, feeling like a burden, or giving up activities we enjoy. But it's time to change our perspective. Instead of viewing these changes solely as problems to solve, we can see them as natural aspects of life. It's essential to treat individuals going through these changes with respect, ensuring they remain part of our community as they are and receive the support they need.

Gabby Giffords, the former U.S. Representative who survived an assassination attempt in 2011, sustained a severe traumatic brain injury from a gunshot wound to the head. She has spoken publicly about her extensive recovery process, emphasizing the difficulty of relearning how to speak — a condition known as *aphasia* — as well as walk and perform other basic tasks.

Giffords, who incorporates speech therapy and music into her wellness plan, told CNN in 2022: "For me, it has been really important to move ahead, to not look back. I hope others are inspired to keep moving forward, no matter what."

REMEMBER

Whether it is due to aging or injury, the thing is, everyone's brain is unique and can change over time. By using a neurodiversity perspective, we recognize that these changes are just another version of normal.

TIP

How do you empower and support these folks? Start by listening, really listening, to their experiences and needs. Make room for them in our communities and in your life, just as they are, not as you think they should be. Ensure they have the tools, resources, and, most important, the respect they deserve to live fulfilling lives.

Thinking About Mental Health

Conditions such as autism or ADHD are part of the brain's structure from the beginning, while others such as OCD or bipolar disorder may develop later due to factors not fully understood. Changes from aging or injury can also alter the brain at any stage.

Mental health is about our feelings, thoughts, and social connections. It's normal for mental health to shift over time. Research has shown that many mental health conditions such as depression and anxiety have genetic components, in addition to environmental, psychological, and biological factors. Many in the neurodiversity movement feel strongly that mental health should be included under the umbrella of neurodivergence.

Neurodivergent people may experience unique mental health challenges that can arise from their brain wiring itself (such as the intrinsic anxiety that can be a part of autism) or from external factors, such as societal misunderstanding, stigma, or the stress of navigating a world designed for neurotypical individuals. For instance, the social and sensory challenges experienced by someone with autism can lead to increased stress or anxiety.

REMEMBER

Though mental health may or may not be influenced by the individual's brain structure, we can adopt the neurodiversity perspective when considering mental health. Recognizing that brain differences are normal allows us to adapt our surroundings, methods, and mindsets to assist those facing mental health challenges, aiding their success.

Where and how trauma intersects

Trauma can deeply affect a person's mental well-being, just as a physical injury impacts the body — often leading to post-traumatic stress disorder (PTSD), anxiety, or depression. For neurodivergent people, this effect may be magnified due to their distinct brain makeup, possibly leading to stronger stress responses. It's key to recognize that one-size-fits-all mental health treatments aren't effective for everyone. Neurodivergent individuals often need customized support to address trauma (see Chapter 24 for more on this).

How to approach mental health

Most people will encounter some form of mental health challenge during their lifetime, whether it's temporary stress, anxiety, a period of depression, or more severe and enduring conditions. It's important to remember that having mental health struggles is a common part of the human experience, and seeking help or support is a good thing.

Addressing stigma

Stigma, the negative judgment of people due to certain traits, frequently prevents people from seeking mental health support. It creates a barrier to reaching out for help or sharing their challenges. To demolish this barrier, mental health should be talked about as openly and unashamedly as are physical ailments like a broken arm. We must understand that mental health issues aren't a choice, a failure, or a weakness.

Actor and producer Viola Davis shared this advice in 2020 on the social networking site X for those who encounter stigma when discussing mental health: "There is no shame in speaking out or seeking help. I myself suffer from anxiety and there is much stigma when you're not perceived as a 'strong' Black woman. But, let's redefine strength as vulnerability, authenticity, and the courage to say 'I'm hurting'."

Mental health issues are normal, and they should be treated as such — by speaking openly and leaving judgment behind. Supporting mental health means listening, educating ourselves, and replacing old stereotypes with true understanding. Everyone's mental health journey is valid.

Applying a neurodiversity frame

TIP

Understanding mental health through a neurodiversity lens is about accepting that all of our brains function differently. Here's how to do that:

>> **Respect:** Avoid judging anyone for how their mind works.

>> **Tailored support:** What helps one person may not help another.

>> **Strengths-based approach:** Focus on the positives, such as the resilience people with mental health challenges often show.

>> **Holistic view:** Mental health is linked to a person's life and experiences, not just their brain.

>> **Teamwork:** People know their own mental health best. Work with them.

Myths and Misconceptions

Many myths and misconceptions are associated with the conditions discussed in this chapter. We dispel a few of the biggest ones here.

People with ID can't learn

It's a myth that people with intellectual disabilities can't learn or have rewarding lives. While they may process information at their own pace, they definitely learn and grow. People with ID also deserve to be a full part of their communities. Everyone has different strengths, and this includes people with ID.

Everyone with Tourette's "swears"

There's a common myth that individuals with Tourette's frequently involuntarily swear. In reality, this is quite rare. Only a small percentage of people with Tourette's have this specific type of vocal tic. Most tics associated with Tourette's are simple, small sounds or movements.

Bipolar means "manic" or "depressed"

The idea that bipolar means just swinging between manic and depressed states is oversimplified. In reality, bipolar involves various moods that differ in intensity and how long they last. People may have times of high energy or intense sadness, but also phases of calm and stability. It's not just flipping between two states.

People with OCD are just "neat freaks"

The notion that OCD is just about being overly tidy is wrong. OCD involves persistent, intrusive thoughts and repetitive behaviors to ease these thoughts. While some with OCD may clean frequently or arrange items precisely, others may fixate on safety concerns or symmetry. OCD isn't simply a preference for neatness; it's a complex experience.

3

Navigating Life as a Neurodivergent Person

Explore whether common neurodivergent traits resonate with you and how these traits can manifest as strengths.

Find out about the diagnostic process and what getting a diagnosis of neurodivergence entails.

Discover why being seen and understood can be challenging for someone with neurodivergence and how neurodivergent individuals can help others to better "get" them.

Find out how to exercise compassionate curiosity and find common ground with those who may not understand you.

Uncover skills you can use to better understand yourself, grow as a person, and connect with others.

Chapter **10**

Understanding Yourself as a Neurodivergent Person

Have you ever wondered why you seem to be so different compared to your family or friends? Why you have so many difficulties with things that others seem to be just fine with? Or agonized over whether something is really wrong with you? If you have, know that you are not alone. Many neurodivergent people have gone through such struggles until they start on the journey of understanding who they really are.

In this chapter, we guide you through the diagnostic process and what getting a diagnosis entails. You also delve into common neurodivergent traits to see whether any resonate with you. Plus, you explore how these traits can manifest as strengths, giving you a more nuanced perspective of what it means to be neurodivergent.

Seeking a Diagnosis

If you're an adult seeking a diagnosis of a neurodivergent condition, you're not alone. It is estimated that neurodivergent conditions may be found in at least one-fifth of the world's population. The following sections walk through the typical steps one takes in exploring a diagnosis. (If you are a parent, see Chapters 19 and 23 for how you can support yourself and your child through the diagnostic process and beyond.)

Understanding the diagnostic process

The diagnostic process for neurodivergent conditions typically involves several steps. It's important to note that the specific process can vary depending on your age, traits, and your health care provider's practices. However, here's a general outline of what the diagnostic process may look like for an adult:

1. **You recognize neurodivergent traits in yourself.**

   If you suspect you may be neurodivergent, you may recognize traits that differ from the "typical," such as difficulty concentrating, challenges in social situations, or specific and intense interests.

2. **You seek information.**

   The next step usually involves personal research. The Internet offers resources and personal accounts that can help you relate to neurodivergent traits. Your research may suggest which specific neurodivergent condition(s) are applicable to you and that may be all that you need for your journey of self-discovery and self-acceptance. Or you may choose to pursue a diagnosis from a professional to confirm your thinking. (If you encounter barriers in seeking a diagnosis, see "Coping when diagnosis isn't possible" later in this chapter.)

3. **You have an initial consultation with a health care provider.**

   The consultation may be with a primary care physician, psychiatrist, psychologist, or neurologist. Usually you will discuss your traits, concerns, and medical history. The provider may also ask about your developmental history and school or work performance.

4. **You are referred for further evaluation.**

   After your initial appointment, your health care provider may recommend that you see a specialist for a thorough assessment. This expert may evaluate such traits as cognition, attention, the way you socialize, and behavior to determine whether they align with a specific neurodivergent condition.

5. **You are offered a diagnosis.**

   After an evaluation, the health care professional will diagnose based on criteria from a diagnostic manual (see Chapter 5 for details). If you are diagnosed with a neurodivergent condition, they'll explain the condition, answer questions, and recommend next steps.

6. **You determine your post-diagnosis needs.**

   After you are diagnosed, the approach varies based on individual needs. Some may pursue coping strategies, connecting with neurodivergent communities, therapy, medications, or deeper self-understanding. Joining support groups, either online or in-person, to share experiences can also be beneficial.

Navigating barriers to diagnosis

Many neurodivergent people encounter challenges in seeking a diagnosis (see Chapter 5 for more). Here are some tips to assist you in the sometimes-lengthy process:

>> **Choose the right professional.** Look for health care providers with experience in diagnosing neurodivergent conditions in adults. Your doctor can often provide referrals.

>> **Bring support.** Having a trusted companion during appointments can offer emotional support and help recall crucial life details you may forget. As Orion Kelly, the Melbourne-based podcaster known as *That Autistic Guy,* says, "If you have a partner, bring them along. They know you better than anyone, and even if you find it hard to remember certain things about yourself, your partner likely won't."

>> **Advocate for yourself.** During the diagnostic process, actively voice concerns, ask questions, and seek clarifications. Ensure your needs are met and views are acknowledged. Be proactive and assertive in your health care journey.

>> **Be patient.** Diagnosing neurodivergent conditions can take time. Be patient and recognize that it's a journey toward self-understanding.

>> **Remember you're still you.** Regardless of an official diagnosis, you remain the same person you've always been. Continue seeking strategies to thrive and be your best self.

REMEMBER

This process can be complex, but it's about understanding yourself and finding support. Persistence and self-advocacy are vital in overcoming challenges.

Coping when diagnosis isn't possible

Limited access to health care is a reality for many. However, if you suspect you're neurodivergent but can't immediately get a diagnosis, there are steps you can take:

>> **Conduct online research.** The Internet offers abundant information. Research your suspected condition on reputable sites or online communities to learn about symptoms, coping methods, and shared experiences. While it doesn't replace professional diagnosis, it can enhance self-awareness.

>> **Join support groups.** Numerous online and offline communities exist for neurodivergent individuals to share experiences, learn, and gain support. Connecting can be immensely beneficial!

>> **Practice self-care and coping strategies.** Even without an official diagnosis, if you're struggling with certain things, consider coping techniques such as getting enough exercise, following a good diet, practicing mindfulness, and getting proper sleep. They can significantly impact your well-being.

>> **Look at public resources.** Explore community health resources. Some offer affordable mental health services. Schools or universities often have counseling services and may provide accommodations for students, even without a formal diagnosis.

>> **Seek workplace accommodations.** If comfortable, discuss your needs with your employer; they may offer accommodations to enhance your work environment.

REMEMBER

Some people, due to difficulty obtaining a diagnosis or other reasons, choose to accept their self-realization as a legitimate condition. This is becoming more accepted in neurodivergent communities. Though a diagnosis is valuable, self-understanding and care are paramount. Stay persistent and find what works best for you.

Profiling Your Neurodivergent Traits

Neurodivergent traits can appear in everyone, including neurotypical people. However, individuals with neurodivergent conditions tend to exhibit large clusters of these traits, making their experiences more pronounced and distinct. If you're reading this book, you may be curious about your own neurodifferences.

This chapter is intended for self-awareness and understanding of traits that differ from most people. It is not a substitute for an official diagnosis and should be used as a starting point for personal insight.

You're on a journey of self-discovery, embracing your unique identity. Understanding your thoughts and interactions doesn't just help you; it offers loved ones a guide to your needs and communication style. It's like giving them a roadmap to build stronger bonds, appreciate your distinctive outlook, and enrich relationships.

Perhaps you have a diagnosis, or maybe you don't. Either way, it's no big deal. A diagnosis is merely a starting point in understanding yourself. Our brains make us all unique, filled with a vast array of thinking and behavioral traits. Some of these may be considered "typical," while others may set you apart from the crowd.

Why bother with all this self-reflection? Well, understanding and accepting your specific traits lay the groundwork for navigating a world that may not always suit the way your brain ticks. So buckle up! In the following sections, we dive deep into exploring these areas to help you feel right at home with yourself.

Heads up! Meeting one neurodivergent person means just that — you've met one individual with unique characteristics and strengths. Remember, being neurodivergent doesn't mean fitting into a one-size-fits-all category. Each person, including yourself, may display some, all, or none of these traits.

Seeing the way you think and learn

When you're chatting with others or trying to understand events, your brain uses a process called *cognition* to help make sense of it all. Think of cognition as your brain's way of learning, thinking, and processing what you experience. This process is influenced by your genes, surroundings, and past events. And just like software updates, it keeps changing as you learn more. For many neurodivergent people, cognition may operate a bit differently than for those who are neurotypical. Take a moment to see whether any of the following cognitive traits resonate with you.

When responding to the following traits, choose how often they apply to you: *never, a few times, half of the time, most of the time, always.*

If you're uncertain, consider asking someone close to you who knows you well. If you select *most of the time* or *always* for any of these traits, it indicates that your experience may differ from what is considered typical.

>> **Decision-making:** You thoughtfully evaluate all options before responding, highlighting your dedication to informed decisions. However, this can also stem from a fear of making the wrong choice.

>> **Literal interpretation:** You often take what people say at face value. If someone makes a commitment, you expect them to honor it. You get sarcasm, but often you are unsure whether a person is really being sarcastic or sincere. You understand things much better when they are given context, as opposed to when they are implied.

>> **Learning:** You like to teach yourself new skills and can think up creative solutions when given enough time and information. You're more likely to examine a new challenge from all possible sides, rather than relying on hunches or past experiences.

>> **Reading:** You take your time when reading things, more so than others. Sometimes you have no problem reading, but other times you may struggle to get through a paragraph.

>> **Estimating:** You have difficulty estimating how much time something may take, how far something is, or how much of something you should buy.

>> **Numbers:** Even when using a calculator, you often get numbers wrong.

>> **Memory:** You remember information from long ago, even when others don't. Remembering things from yesterday or this past week can be more difficult.

>> **Doing things:** You prefer doing the same task repeatedly and have difficulty determining when to move on to the next task. When doing a project, you like to refine it again and again until you really get it right.

>> **Money:** You have difficulty keeping budgets, or you wonder where all the money went.

>> **Focus:** Your ability to pay attention to a task you like doing is much stronger than most. Being interrupted in the middle of it really throws off your concentration.

>> **Intense interests:** You spend a lot of energy on your deep interests, which may lead to neglect of other areas.

>> **Stress:** When stressed, you feel the need to flee and hide. Or perhaps you've noticed that you feel a much stronger emotional response to stress than most.

You may worry that these traits are judging you, but let's set the record straight: We're not. None of these observations mean that you are "broken." Instead, they simply highlight that having differences can bring both unique strengths and challenges. No person has it totally easy in life.

Consider this list as a guide to common cognitive traits. It's not universal; these traits can vary widely. Whether some, all, or none apply to you, it's a starting point to understand your very own unique brain.

Decoding how you communicate

Everyone communicates, but not all in the same way. If you're neurodivergent, your communication style may be unique compared to neurotypical folks. By getting to know your own way of communicating, you're on the path to better connections with others.

Use the following list to map out your communication profile, using the same instructions from the previous section:

>> **Speaking:** You experience difficulty communicating by speaking and participating in conversations. Maybe you excitedly cut-in when others are talking, quickly switch topics, hold back because you're unsure when it is your time to speak, or repeat words like a chatty chorus.

>> **Reciprocity:** You love to talk about your interests, but sometimes forget to let others talk about theirs. You sometimes wonder whether you forgot to let the other person speak.

>> **Speaking style:** Your answers to questions tend to be super short or extra-long. Some folks think you share too little or too much.

>> **AAC devices:** You use iPads, notetaking apps, or use an augmentative and alternative communication (AAC) device to talk instead of using vocal speech.

>> **Switching topics:** When the conversation moves away from a favorite topic, it's tough. You may feel a pull to return and finish what was left unsaid.

>> **Writing:** You express your thoughts better through writing than speaking. Perhaps you send long text and emails while others send short notes. Alternatively, perhaps you dread a bombardment of texts and prefer to chat over video or face-to-face.

REMEMBER

Figuring out these traits opens doors to easier communication. If you find writing easier, go ahead and use it to express yourself. Trouble with conversation? Learn about how others use social cues and turn-taking. Do you tend to go on too long? Ask people to interrupt you and ask for their turn, rather than simply roll their eyes if you get too excited or forget. It's really about using what you're good at and using self-advocacy to explain and advocate for your social communication style.

Getting to know how you socialize

Recognizing how you engage with others in social situations can shed light on your neurodivergent traits. Socializing is something everyone does, but some people approach it in unique ways that differ from the majority.

Use the following list to profile your social interactions, using the same instructions from the previous section:

>> **Eye contact:** Eye contact may feel uncomfortable or intense for you. Instead, you find yourself looking elsewhere, such as at a person's mouth or nose, during conversation.

>> **Anxiety in new encounters:** You feel anxious when meeting new people.

>> **Crowds:** Being in crowded places or unfamiliar gatherings can overwhelm you. If you know only one person at an event, you tend to stick close to them.

>> **Sharing interests:** You don't enjoy small talk ("Why are we talking about the weather?!"), but when you find that you share an interest with someone, you can talk for hours.

>> **Social interactions:** You prefer deliberate reasoning over intuition in making sense of social situations.

>> **Understanding others:** You're unsure when others are being fake and don't always know whether someone is being sincere. You wish people would act and say things according to how they really feel.

>> **Tipping:** You tend to round up when tipping, even if a smaller percentage is expected. Splitting the bill at dinner with friends can produce a bit of anxiety.

>> **Energy drain:** Social interactions tend to drain your energy. You need time alone to recharge.

>> **Understanding reactions:** You sometimes don't understand why people are upset with you.

>> **Facial expressions:** People often assume you're upset, even when you're completely happy.

>> **Remembering:** You may easily recall detailed information about a person but struggle to remember their name or face without a bit of context.

>> **Unspoken rules:** You find it difficult to figure out unstated rules at work or in social situations.

Knowing your traits is a big step toward shaping how you interact with others. If eye contact makes you uncomfortable, you may find other ways to show you're listening. Recognizing what tires you out can help you set limits that keep you comfortable. Whether it's using your strong interest-based social skills or finding ways to manage challenges, it's all about understanding yourself and using that knowledge to your advantage.

Learning how you get things done

Executive function is like your brain's manager, helping you plan, focus, and carry out tasks. We refer to it as "how you get things done." Ever had things stray from your plan? The skill to stay on track despite hiccups, while controlling emotions, falls under executive function. It's key for succeeding in school, work, and daily life. For those who are neurodivergent, understanding how you handle these functions can highlight strengths and areas for improvement.

Use the following list to map out how you get things done, using the same instructions from the previous section:

>> **Impulse:** You notice you have a stronger impulse than most, and you often act without thinking. This can be great when at parties or in needing to make quick decisions, but this impulse sometimes leads to later regretting the consequences of your actions.

>> **Intuition:** You like to reason things out rather than just relying on a hunch.

>> **Attention:** You struggle to concentrate.

>> **Flexibility:** When you're in the "flow," or really concentrating on something, you can have difficulty switching between tasks.

>> **Task management:** You struggle with ordering tasks correctly, or thinking of alternative solutions when how you're doing something just isn't working.

>> **Time management:** You tend to start tasks at the last minute, misjudge their difficulty, or have trouble prioritizing when faced with multiple tasks.

>> **Grasping time:** You often get dates and times wrong, routinely show up to things too early or late, or struggle with translating between time zones.

>> **Help:** It's hard for you to ask for help. You prefer to figure things out on your own.

>> **Monitoring progress:** You struggle to track your progress toward a goal, make adjustments as needed, or apply past experiences to new problems.

>> **Emotions:** You struggle to deal with unexpected events calmly and deliberately. You strongly react to minor problems, easily feel hurt, or quickly become upset.

- **Organizing:** You struggle with multitasking, organizing things, planning, and locating items.

- **Initiating tasks:** You have a difficult time starting tasks that need to get done, but when you are faced with a deadline, you have no trouble getting it done.

- **Verbal instructions:** You have difficulty remembering verbal instructions for complex tasks.

- **Written instructions:** Following a series of written instructions is difficult. You prefer to watch a video explaining how to do something than read an article or manual explaining how.

- **Big picture:** You lose sight of the overall goal while focusing on the details. Alternatively, you're great at "big picture" things but have difficulty paying attention to the details.

These are just broad areas of differences. Not all neurodivergent people experience these things, and few experience all of them.

Exploring how you experience the world

If you're neurodivergent, you may sense the world differently than others. This can have a big or small impact on your daily life, and you may need various levels of support or adjustments to manage these differences. One common way to cope with sensory input is through self-stimulation, often referred to as *stimming*. Stimming is something all people do to manage stimuli. Stimming includes doodling, tapping your hand, twirling your hair, rocking your body, pacing, repeating words or phrases, and swiftly flipping through online content to find calm, to name a few.

Use the following list to map out how you process the world through your senses, using the same instructions from the previous section:

- **Sound:** Sounds can be a big deal for you. You may be sensitive to the noise in bustling environments like coffee shops or concerts, or even to particular sounds like sirens or vacuum cleaners. While these noises may annoy or discomfort others, sounds may even cause pain. On the other hand, your heightened sensitivity may allow you to pick up on subtle sounds like an insect buzzing.

- **Light and color:** Bright lights, flickering lights, and certain colors or patterns bother you. They may even bring on migraines or seizures.

- **Temperature:** You often feel too hot or too cold in most places.

- >> **Touch:** You can be sensitive to touch sensations such as hugs, handshakes, getting your hands wet, or the feel of clothes on your skin.

- >> **Personal space:** You are highly sensitive to others entering your personal space.

- >> **Food or smell sensitivities:** You have heightened sensitivities to the taste, texture, or smell of food. You are highly sensitive to certain smells.

- >> **Pain tolerance:** You can tolerate pain more than most people, and those "Rate your pain on a scale from 1 to 10" questionnaires just don't resonate with you. On the other hand, you may be more sensitive to pain than others around you.

- >> **Stimming:** You need to stim more than most.

- >> **Information input:** You need constant stimulation, or you easily become bored.

- >> **Body awareness:** You may not always be aware of the positions and movements of your body.

REMEMBER

Understanding your neurodivergent traits helps you adapt to your surroundings. For example, if loud noises cause you to avoid certain places, noise-cancelling headphones may help. Recognizing your need to stim can lead to personal calming techniques. By tuning into these traits, you can turn challenges into opportunities for comfort and success.

Uncovering your strengths

Everyone has strengths and challenges, including those with neurodivergent traits. While not superpowers, these unique strengths can stand out and be embraced, allowing you to celebrate your particular abilities and perspectives. Check out the following strengths to see whether any are familiar to you:

- >> **Attention to detail:** Your strong attention to detail enables you to excel in tasks needing precision and accuracy. This knack for details can be a great asset in many fields.

- >> **Creativity and innovation:** You think outside the box and approach problems in unique and innovative ways, leading to creative solutions and breakthroughs.

- >> **Hyperfocus:** You can hyperfocus on tasks that interest you, allowing you to accomplish a lot.

- >> **Pattern recognition:** You excel at recognizing patterns, using them to solve complex problems.

- » **Intense interests:** You have passionate interests in specific topics or activities and can easily learn everything you can about them.

- » **Strong memory:** You have strong memories and can recall details and information that others may forget. You're often great at games of trivia.

- » **Empathy:** You have a deep understanding of the emotions of others. You're the "go to" person when things are tough. You listen and are an excellent caregiver.

- » **Integrity:** You tend to be more honest than most. You keep your commitments and your word.

- » **Resilience:** You have a strong ability to bounce back from challenges.

REMEMBER

The strengths mentioned here are often found in neurodivergent individuals, but they can vary widely. You may have some, all, or none of these traits, and that's perfectly fine. The key is to discover your unique abilities and build on them, embracing what makes you YOU!

Finding Success

Success starts with understanding how you're like others and how you're different. Your foundation is self-awareness. Your differences aren't defects; they're just part of who you are. Embrace them, and build on this foundation of self-awareness to accept yourself as you are, without feeling broken or needing to be fixed.

WARNING

Self-acceptance doesn't mean you shouldn't work on improving skills; everyone needs to!

From understanding and accepting yourself grows *self-confidence*. Self-confidence about recognizing your unique strengths and areas where you need support. Whether you're thinking about how to apply your strengths in social, educational, or work settings, or identifying ways to grow, self-confidence acts as a guide. It helps you to let your abilities shine and figure out how to use your differences to your advantage.

Chapter **11**

Understanding Others

Every human craves recognition. We all want to be seen and understood, and we desire to have our way of seeing things acknowledged and appreciated by others. However, this pursuit can be challenging for neurodivergent individuals as historically, society struggles to comprehend our experiences and often ignores us or tries to fix us. This lack of understanding can create barriers and stigmas, which often hold us back.

So, if you're a neurodivergent person who wants others to get you, how do you do that? Well, it includes deepening your self-knowledge (discussed in Chapter 10) and finding practices that help you thrive (discussed in Chapter 13). Another key aspect is striving to understand those *who don't understand you*. It may seem contradictory at first glance, but stay with us. In this chapter, we show you how.

Recognizing How Others Think and Experience the World

We all have our own ways of doing things. Even on our author team, each of us has a different approach to writing, editing, and organizing things. If you're neurodivergent, you have a unique perspective to bring to the table. But it's important to remember that other people — neurodivergent or not — also have their own unique ways of looking at and understanding things.

Remembering this isn't always easy. You see, our brains are hardwired to think of ourselves as the center of the universe. It's not egotism, exactly — it's more like an evolutionary instinct to prioritize our own survival. However, it often causes us to forget that others have experiences just as rich, complex, and varied as our own.

Understanding others' perspectives offers benefits. First, it fosters empathy. Seeing from another's viewpoint helps comprehend their feelings and needs. Second, it improves communication. Knowing how others perceive and process information allows for more effective ways of discussing things.

There's a benefit, too, beyond just the personal level. A society in which people take the time to understand one another allows them to work together more effectively to avoid conflict and solve big problems. It's a world that's less divided, more compassionate, and more resilient in the face of adversity.

REMEMBER

It's not just about you; it's also about *them*. When you validate someone else's experiences, you make them feel heard and understood. That's a powerful thing — it can build trust, strengthen relationships, and even help to soothe pain and discomfort. And you may just learn something new along the way.

Consider the scenario where three friends attempt to simultaneously pass through a slim doorway. In doing so, they inadvertently mash their bodies, creating a jam. Only moments before, they were happily approaching the entrance together. However, by rushing to push their individual selves forward, it causes gridlock.

You can easily apply that analogy to life. People often rush headlong into the doorways of our own lives, eager to press on with our own individual narratives, seeking to be recognized and understood. That's natural, and we get that! But, there's lots of other people doing the same — in fact, *every other human* is often doing so. That's a lot of bodies in a lot of doorways, all seeking to be seen and to be heard all at once.

TIP

There's a rather simple solution to everyone everywhere wanting to be seen and heard all at once: Pull back, pause, and allow the other person to be heard first.

"Well, wait," you may say. "If I want to be understood, why should I encourage others to share their views first?" It's not only the kind thing to do, but it's also a good strategy in seeking to be understood.

Imagine there are two roommates, Alejandro and Ben. Alejandro comes home after a long day to find the kitchen sink full of dirty dishes — again — all courtesy of Ben. His instinct is to storm up to Ben and demand that he wash them. But he remembers the life lesson of seeking to understand the other person first. Alejandro pauses and realizes that understanding is a two-way street. He decides to swap roles and mentally steps into Ben's shoes, peeling back the layers of his own frustration, and asks, "Why?"

Why does Ben consistently leave dishes in the sink? Is he overwhelmed with work? Is he simply unaware of the impact and stress that his actions have on Alejandro? By seeking to understand before being understood, Alejandro opens the door to a conversation rather than a confrontation. This mindful approach is what allows shared understanding and *change*.

Another example is a certain segment of autistic children who exhibit behaviors that may seem aggressive to some. Again, strive to understand. Living in a world that largely doesn't understand or accommodate them, these children often struggle to communicate their needs, emotions, or discomfort over things ranging from overwhelming sensory experiences to difficulty with changes in routine. As with every other human, they're just struggling to be seen.

Practicing Compassionate Curiosity

When it comes to understanding others, we're a big fan of what we like to call *compassionate curiosity*. We frequently introduce this phrase to neurotypical individuals, encouraging them to practice it to reflect on the experiences of the neurodivergent people they interact with. However, the concept of compassionate curiosity is just as applicable to neurodivergent individuals as well.

So, what is compassionate curiosity? It's a state of mind where one approaches interactions with others from a place of both empathy and genuine interest. It involves the willingness to engage in open, nonjudgmental exploration of another's experiences, emotions, and perspectives and comprises two crucial components: compassion and curiosity.

>> **Compassion** is a deep awareness of and empathy toward another's suffering, joy, struggles, and success (or whatever state they may find themselves in). In the context of compassionate curiosity, compassion refers to approaching conversations and relationships with a sense of kindness and understanding, seeking to grasp the other person's emotional state and experiences without diminishing or dismissing them.

>> **Curiosity** is the desire to learn or know about something. When you are curious, you're open to new ideas, actively seeking to understand rather than passively accepting what's presented to you. In this context, curiosity means genuinely wanting to understand the other person's point of view, their experiences, and their feelings.

Combining these two elements, *compassionate curiosity* means striving to understand others, not just for the sake of knowledge, but with the intention of empathizing with their experiences, acknowledging their feelings, and fostering a connection based on mutual respect and understanding.

Practicing compassionate curiosity can help you communicate better, reduce conflicts, strengthen relationships, foster personal growth, and make your needs known. It involves active listening, asking thoughtful questions, and being open to perspectives that may differ from your own. It's about creating a space where open, honest, and empathetic conversation can occur.

When you take the time to understand someone else's perspective, it can help you see things in a new light. Maybe they have a solution to a problem you're facing that you never would've thought of on your own. Maybe they can help you understand a concept that you've been struggling with. Plus, when you understand where someone else is coming from, you can communicate with them more effectively, leading to more productive discussions and less conflict.

First things first: Seek to understand

Remember the earlier example of Alejandro and Ben and the issues they had around dirty dishes? Well, that's an example of how exercising your compassionate curiosity starts with striving to understand the other person first. Just as in constructing a building, understanding serves as the foundation of any relationship. Without this crucial base, every other component — all the exchanges, shared moments, commonalities, and differences — risks crumbling down. Sure, superficial interactions may suffice temporarily, but for a lasting, stable relationship, a solid foundation of understanding the other person is crucial.

Now, we're not talking about a surface-level understanding. This is not about simply giving a nod of the head and a sympathetic "Oh, I see." No, we're talking about deep, genuine understanding. The kind that takes time, effort, and patience. The kind that comes when you really, truly listen to another person, when you put aside your own assumptions and prejudices and open your mind to their experiences.

Here's an example: Jaime is dyslexic. When Jamie reads, the process Jamie follows is different. Words and letters may jumble, making it a bit harder for Jamie to decode text. But that doesn't mean Jamie can't understand complex ideas or perspectives. It's not about intellect; it's about how Jamie processes information.

Now suppose Jamie's chatting with Emily, who's neurotypical. Emily is sharing a love for books, for losing oneself in the written word. Now, Jamie could brush this off and say "Well, books just aren't my thing," but instead, Jamie decides to practice some compassionate curiosity.

Jamie asks, "What's it like for you when you read? What kind of books do you enjoy?" Jamie listens, not just to understand Emily's love of reading, but also to understand Emily. And Jamie shares, too. "When I read, it's a little different because of my dyslexia. But I love stories told in other ways, like through movies or podcasts."

It's not about changing who you are to match someone else, or even about overcoming differences. It's about understanding your differences, learning from them, and finding common ground. That's how Jamie builds a bridge with Emily, not by pretending to be someone else, but by seeking to understand and being open about their own experience, too. And that's the beauty of compassionate curiosity, right there.

When you seek to understand, you're saying, "Hey, your experiences, your feelings, your perspective? *They matter.* They're important. And I want to know more about them." It's a way of validating the other person, showing them empathy and respect.

REMEMBER

Genuinely seeking to understand others often encourages them to better understand you. However, remember, *this isn't always so.* Some people may not seek to understand you, *and that's okay.* Your focus should be on understanding them through compassionate curiosity. What they do is their choice. You can only manage your own actions.

TIP

Here are some ways you can genuinely seek to understand the other person:

>> **Listen actively:** Active listening can be a bit of a buzz term, but at its heart it's about really listening to the points, feelings, experiences, and message of the other person.

>> **Hold up:** When others are talking, many of us are already thinking about the next thing *we're* going to say. Stop. Just listen to the other person without pre-forming your own response.

>> **Ask insightful questions:** Many neurodivergent individuals naturally discern patterns, making them excel here. Use questions to delve deeper and better comprehend others' viewpoints.

>> **Don't fix:** This can be challenging, especially for individuals with autism or ADHD, as their brains can often identify solutions quickly. Yet, people mostly want understanding. If you do want to propose a solution, inquire whether that's what they want. Often, yes! Other times, no.

- **Put assumptions on the shelf:** Our brains love to fill in gaps with assumptions. It's natural, but not always helpful. Try to put those aside and stay open to what the person is actually saying.

- **Be patient:** It seems "be patient" is one of our most frequently offered tips, but that's because it works. Understanding doesn't come instantly. It's a process. Be patient with others and with yourself.

- **Reflect their feelings:** Okay, so now you've *heard* them. Now, show them that you get it. Repeating a short summary of what they said is a good way to demonstrate that you're really tuning in.

- **Stay respectful:** Even when you disagree, you can still understand their perspective. Disagreements don't have to be roadblocks to understanding.

Build bridges

Now you have a better understanding of the other person. Great! That doesn't mean that you see the world from the same perspective or that you 100 percent agree on everything. But, you know what? *That's fine.*

TIP

Healthy relationships aren't about changing those around you. Instead, they're about fostering mutual understanding, practicing open communication, demonstrating support, and growing and learning alongside others with whom we share this planet. Here are some ways you can build bridges with others:

- **Acknowledge differences:** Spot them, accept them, and understand differences add color to our world.

- **Find common ground:** Look for shared experiences, interests, and feelings.

- **Communicate respectfully:** Share your views, but remember that others have their views too.

- **Practice compassionate curiosity:** Right back where we started! Yes, compassionate curiosity is a *practice*. Keep understanding, keep asking, keep listening.

Chapter **12**

Improving Personal Effectiveness

I n this chapter, we look at skills you can use to better understand yourself, grow as a person, and connect with others. It is *not* about changing who you are. Rather, it's about helping you be authentically you.

We start by first exploring trauma. Why? Well, many of us neurodivergent people encounter trauma by living in a world not designed for us. Next, we discuss the type of mindset that can help a neurodivergent person thrive in life. We then address personal effectiveness — those habits and practices that can help you achieve your goals. We also share practical tips on how to live an authentic, resilient life.

This chapter is not about changing who you are as a neurodivergent person. It's about helping you thrive in a rather neurotypically focused world.

REMEMBER

Understanding the Impact of Trauma

Trauma is a distressing or life-threatening experience — such as an accident, loss, or act of violence — that causes a deep emotional wound. Trauma can profoundly affect a person's mental, physical, and emotional wellbeing. Bullying, abandonment, rejection, and other negative social interactions can create trauma as well. When you experience trauma, it can shake up your world, leaving you feeling overwhelmed, scared, or alone, much like being stranded in a storm without an umbrella.

What's trauma like for neurodivergent individuals? Some may endure recurring feelings of discomfort, anxiety, or depression without understanding why. Trauma can feel as though a smoke alarm is sounding, but the smoke source isn't clear. They may react quite strongly to situations that subconsciously remind them of the trauma, much like tripping over unseen furniture. Trauma's manifestations are complex and unique to each individual, underscoring the crucial role of mental health professionals in addressing trauma. This is true for both neurodivergent and neurotypical individuals.

REMEMBER

Some folks can be completely oblivious to the fact that they've experienced trauma. You see, our minds are amazing and tricky all at once. Sometimes as a defense mechanism, the mind suppresses or tucks away traumatic experiences. This is its way of trying to protect us from pain.

If unaware of their trauma, individuals may feel as though they're on an emotional roller coaster. Intense reactions to minor triggers such as loud noises or sudden movements can occur as the fight or flight response overreacts. Nightmares, anxiety, unexplained feelings of sadness, anger, or fear, along with physical symptoms such as headaches, fatigue, or a chronic sense of being unwell may emerge. They can feel as though they are living in a mystery novel, puzzling over their feelings.

REMEMBER

Recognizing trauma is challenging, but with the right support, you can uncover these hidden wounds if you have them and embark on the path to healing.

Trauma and PTSD

About 20 percent of people who experience trauma develop post-traumatic stress disorder (PTSD). PTSD is a mental health condition characterized by heightened arousal, anxiety, intrusive memories, and avoidance of anything associated with traumatic events.

You can think of trauma as a bad injury to your emotions — kind of like a deep cut to your soul. PTSD, then, is like an infection that sets in after that emotional wound. It encompasses a specific set of reactions and symptoms that can develop after trauma. PTSD overwhelms an individual's ability to cope and may cause feelings of helplessness, diminishing their sense of self and their ability to feel a full range of emotions and experiences.

Trauma and neurodivergence

Why discuss trauma in a book about neurodiversity? The link between trauma and neurodivergence may not be initially clear, but there's considerable overlap. Neurodivergent individuals often face unique challenges in a world primarily designed for neurotypicals, leading to increased trauma risk. In addition, trauma can affect the brain — possibly more significantly in those with unique brain wiring. Hence, understanding trauma helps those of us with neurodivergence better support ourselves by addressing our unique needs.

Identifying sources of trauma

In the neurodivergent person, direct trauma may come from bullying or mistreatment due to being different, while indirect trauma may stem from the chronic stress of trying to fit in or conform to neurotypical standards. In some cases, the struggle to get a proper diagnosis and support can also be a traumatic process.

Here's the hopeful part: Just as your brain changes in response to trauma, it can also heal. This is thanks to a wonderful thing called *neuroplasticity*. Your brain can adapt to a sense of safety and healing. We talk more about this in the section, "Healing from trauma" later in this chapter.

Recognizing the impact of trauma

Trauma can negatively impact your personal effectiveness, the set of abilities and traits that allow you to achieve your goals. Here are some examples of trauma's negative impacts:

>> **Thinking skills:** Trauma can mess with your concentration, memory, and decision-making abilities. You may struggle with forgetfulness, have trouble focusing, or have difficulty making choices. This can make it harder to be productive at work or school.

>> **Emotions:** Trauma often interferes with how you handle your feelings. You may experience sudden mood swings, intense anxiety, or angry outbursts, which can affect your interactions with others and your ability to handle stress.

>> **Physical health:** It's not unusual for trauma to cause physical problems such as fatigue, sleep issues, or even chronic pain. These physical symptoms can make it tough to function well in your daily life, including working or completing personal tasks.

>> **Relationships:** Trauma can make it hard to build and maintain healthy relationships. You may struggle with trust, fear of closeness, or find it tough to understand and express emotions. This can strain both personal and professional relationships.

>> **Self-perception:** Trauma can eat away at your self-esteem and confidence. It may make you doubt your abilities, leading to decreased performance and effectiveness in completing various tasks.

>> **Unhelpful coping strategies:** Sometimes you may turn to unhealthy ways of coping, such as substance abuse or self-harm, and such habits can get in the way of functioning well in different areas of your life.

REMEMBER

When it comes to trauma, people react differently. But the effects of trauma don't have to stick around forever. With the right support, individuals can heal from trauma and get back to their awesome, effective selves.

Healing from trauma

After you've experienced trauma, your brain physically changes in response. It's not that your brain is damaged; rather, it's adapted to protect you from danger. Your brain becomes extra vigilant to help you avoid threats. The brain is incredibly adaptable and rewires itself based on what you experience.

TIP

Healing from trauma can seem like a massive mountain to climb, but with some patience, understanding, and good old self-care, it's achievable. Here's a friendly rundown of some things you can do to help yourself heal if you have experienced trauma:

>> **Establish a routine:** Create a daily routine to bring stability and predictability to your life. Set a regular wake-up time and regular mealtimes and make time for activities you enjoy.

>> **Take care of your body:** Your physical health affects your mental well-being. Eat balanced meals, get enough sleep, stay active, and avoid substances that worsen stress.

>> **Practice relaxation:** Try mindfulness, deep breathing, yoga, or meditation to reduce stress and stay grounded. Simple stretching and deep breathing can make a big difference.

>> **Connect with others:** Reach out to trusted friends or family for support and companionship. Sharing your experiences is optional, but the company and distraction can be beneficial.

>> **Express yourself:** Talk about your experiences when you're ready, or find other outlets such as journaling, art, or music to express your emotions and avoid bottling them up.

>> **Stay mindful:** Practice mindfulness to stay in the present and avoid getting stuck in traumatic memories. Focus on your breathing, body sensations, and the world around you.

>> **Use positive self-talk:** Combat self-doubt and negative thoughts by reminding yourself of your strengths and progress. This may sound a bit cliché, but it's really helpful.

>> **Seek professional help:** Don't hesitate to reach out to therapists or counselors trained in trauma healing. They can guide you through therapies such as cognitive-behavioral therapy (CBT), eye movement desensitization and reprocessing (EMDR), trauma-focused cognitive-behavioral therapy (TF-CBT), and somatic experiencing therapy. These therapies all aim to restore your inner sense of safety and clarity.

REMEMBER

Everyone's journey is different. Find what works best for you and don't rush yourself. Healing takes time, but with patience and persistence, you'll get there. And remember, it's completely okay to ask for help!

Choosing a Productive Mindset

To cultivate your personal productivity, you need to start with your habitual thinking patterns. These thinking patterns are called a *mindset*. A mindset refers to a set of beliefs or a way of thinking that determines one's behavior, outlook, and mental attitude. It's like a mental lens through which you view the world. Your mindset plays a critical role in how you interpret and react to situations, set goals, and whether you view obstacles and challenges as opportunities or threats.

It's helpful to think of mindsets like the settings on your smartphone. You know how you can adjust the brightness, sound, or switch between light and dark modes? Well, a mindset is like those settings, but for your brain. It's a bunch of beliefs that shape how your mental operating system interacts with the world.

When it comes to types of mindsets, imagine that your brain had a growth mode or a fixed mode. Psychologist Carol Dweck coined these terms. Someone in a fixed mode believes that their talents, smarts, and skills are pre-set and unchangeable, kind of like how a turtle can't change its shell. But switch over to a growth mode and suddenly your brain believes it can learn and improve, much like leveling up in a video game.

REMEMBER

Your mindset is the software guiding how you think, act, and feel about everything from learning something new to how you bounce back after life throws you a curveball. It's what shapes your actions and attitudes toward life.

Looking at fixed versus growth mindsets

Whether or not you have a fixed or a growth mindset makes a world of difference in how you handle life's ups and downs. When your mindset is in growth mode, it's as though you're driving on a wide-open road — you're resilient, eager to learn, and see each challenge as an opportunity to level up. However, if you're in fixed mode, it's like being stuck in bumper-to-bumper traffic. You may believe you've maxed out your abilities, you may avoid challenges such as unexpected roadblocks, and that fear of failure? It honks at you like an impatient driver.

Your mindset also influences how you handle *challenges*. If you're in fixed mode, challenges may appear as towering monsters, threatening your belief in your natural abilities. On the flip side, if you're rocking a growth mindset, you see challenges as exciting puzzles to solve. They're chances to learn and grow, even though they're tough.

Now, how about *obstacles?* Think of challenges as an uneven, gravel road that jostles and shakes you while you're on it. An obstacle is like a road sign that tells you to "STOP" or "Turn Around!" It's easy to give up when you encounter an obstacle. Even if you know that you have the right to go forward and you *want* to go forward, sometimes dealing with the hassle of an obstacle doesn't feel worth it. Applying a growth mindset reveals that obstacle for what it really is — something to deal with, figure out, and get past or around where possible.

Your mindset also influences your *effort* — that energy you pour into getting stuff done. With a fixed mindset, you may see effort as pointless. If you're not good at something from the get-go, why bother, right? But in the growth mindset corner, effort is a valued resource. It's seen as the key to unlocking your potential and improving your skills. It's all about the grind and the growth.

Nobody likes *criticism,* even when done lovingly it can still feel bad. It would be amazing if we could avoid it. But you know what? We can't. It's a part of life. All we can do is control our reaction to it. A fixed mindset experiences criticism as something personal, and the body feels it as an attack. This can drum up loads of anxiety or shame when this happens. But you deserve to live without tension, panic, and shame! A growth mindset enables you to encounter criticism, pick it apart, examine it, and then learn, adapt, and improve. That feels so much better!

Last but not least is *jealousy*. While feelings of jealousy are often thought of as "wanting what someone else has," deep down it is a fear of losing something that you value. In jealousy, you may experience the success of others as a threat to your own self-worth. A fixed mindset is what drives you off course and finds you crashing into jealousy. A growth mindset enables you to keep a steady hand on the wheel and just cruise along. A growth mindset helps you see clearly that not everything is about you, that the success of others can be inspiring, and that it's a chance to learn from them and maybe even replicate their success.

Adopting a growth mindset

Whether you have a fixed or growth mindset can really shape your journey, be it personal or professional. Having a growth mindset means you see challenges as opportunities to learn and mistakes as valuable feedback. It can boost your personal life, making you more resilient and adaptable. In your professional life, it can foster innovation, problem-solving, and continuous learning, setting you up for success in an ever-changing world.

REMEMBER

Mindsets aren't set in stone. With a bit of self-awareness and work, anyone can shift from a fixed to a growth mindset. Trust us; it's a switch that's well worth the effort! Recognizing the existence of your fixed mindset is a great start. You're now in a position to challenge it and toggle it to growth mode.

TIP

How do you switch on a growth mindset? Here are some steps you can take:

>> **Listen for that pesky fixed mindset voice.** It may chatter things such as "You can't do this" or "You failed again!" Remember, it's just a voice, not reality.

>> **When you hear that voice, pay attention to it.** Catching this voice is a good thing. Most often, when we act on what a fixed mindset is telling us, we do so without really thinking about it. By catching that fixed mindset voice, you're already inching toward a growth mindset. You've spotted the choice — you don't have to buy what the fixed mindset is selling.

>> **Challenge that fixed mindset with your growth voice.** If the fixed voice says, "You're no good at this," counter with, "I'm not good yet, but with practice, I'll improve."

>> **Do what your growth mindset says to do.** This is very important. Otherwise, you're just feeding your fixed mindset. Take the next steps to move forward!

See how that works? To paraphrase Missy Elliott, an American singer and record producer: You put your thing down, flipped it and reversed it. Now you're in a position to grow.

REMEMBER

It turns out the mindset with which you deal with trauma has a huge impact on healing. With a fixed mindset, you view those experiences as pervasive, permanent, and something you can do nothing about. With a fixed mindset you end up in a state of *learned helplessness*. On the other hand, if you choose the growth mindset, you see these experiences as specific, impermanent, and something you can learn from and overcome. With a growth mindset you develop a state of *learned optimism*.

Cultivating Habits of Effectiveness

Habits are the small decisions and actions you perform daily, almost without thinking. They're pretty powerful, shaping your actions, decisions, and ultimately, your life. Over time, habits can significantly impact your health, productivity, and happiness.

Habits can save cognitive resources because the brain doesn't need to exert much effort to carry out these routine actions or behaviors. They can be powerful tools in shaping the kind of life you want to live. If you create positive habits, such as exercising regularly or following a healthy diet, you can greatly improve your overall well-being and satisfaction with life.

However, habits can also be harmful if they lead to behaviors such as smoking or overeating. So how do you develop positive habits and curb harmful ones? One helpful way is by looking at habit formation in the context of the *habit loop*. The habit loop is a concept presented by Charles Duhigg in his book, *The Power of Habit: Why We Do What We Do in Life and Business* (Random House). It provides a framework for understanding how habits work and how they can be changed. It consists of three main components:

>> **Cue:** This is the trigger that initiates habitual behavior. It can be a location, time of day, people, action, or a specific emotional state. For example, your alarm ringing can be a cue to start your morning routine.

>> **Routine:** This is the behavior or action you perform in response to the cue. For example, after hearing your alarm ring, your morning routine may involve brushing your teeth, taking a shower, making coffee, and so on.

» **Reward:** This is the benefit or positive feeling you get after performing the routine. It serves to reinforce the habit, making your brain want to repeat the routine in the future when the same cue is encountered. In the morning routine example, the reward may be the taste of your coffee or the feeling of alertness after your shower.

Understanding the habit loop helps break negative patterns and form healthier ones. When you recognize the cue and reward of a bad habit and *change* the routine, this is known as *habit replacement.* For example, swap stress-induced unhealthy eating for a walk or practicing mindfulness. Over time this can lead to the formation of healthier habits.

Valuing yourself

Your most precious resource is you. But defining *you* is complex. Are you merely the body? The mind? What about your heart, your spirit? Well, you're actually an incredible, interconnected mix of these elements, amounting to much more than their sum. To be truly effective, it's vital to cultivate habits that nurture and elevate your whole self — body, mind, heart, and spirit.

Preserving and enhancing your body

In this section, we chat about how to take good care of your body. We aren't talking about physical perfection or being in the same shape as a blockbuster action star (they have personal chefs, dedicated trainers, and contractual requirements to help them do that). It's more like maintaining a car. If you want it to run smoothly, you've got to put in the right kind of fuel and give it regular tune-ups.

TIP

When it comes to physically maintaining yourself, here are some habits that we find quite helpful:

» **Nutrition:** Think of food as your fuel. Aim for a balanced diet, one that gives your body all the nutrients it needs. Your body will run better, and you'll have more energy.

» **Sleep:** We can't stress this one enough. Just as a car needs to be switched off and parked for a while to cool down, your body needs sleep to recover and rejuvenate.

» **Exercise:** You can think of this as a tune-up for your body. Regular workouts help you build strength and improve your stamina. Move your body in ways that you enjoy and that keep you fit and healthy.

Preserving and enhancing your mind

Just like your body, your mind can be sustained and enhanced through positive habits. Here are a few to explore:

>> **Reducing stress:** Meditation, quiet walks, hobbies, reading, or writing down your thoughts are some of the ways to hit the reset button on your body's stress levels. They can release tension, help you relax, and improve your focus and mental clarity.

>> **Switching on growth mode:** With a growth mindset, you believe that you can improve and grow, rather than being perpetually stuck at where you are now.

>> **Learning new skills:** The more skills you have, the more problems you can solve. Plus, it can be very enjoyable to learn and implement a new skill.

>> **Teaching what you know:** Sharing your knowledge and skills with others is a great way to reinforce your own learning, and it's a wonderful feeling to see those around you benefit from your experience.

Preserving and enhancing your heart and spirit

Next up, we look at how you can fuel your emotional core, the heart. We're not talking whole grains and jogging — though they're important too! Extending beyond your emotional core is your spirit — the essence of everything you. It is where you feel connected to something greater than yourself, such as your community, your family, your faith, or the world all around you. Here are some ways you can nurture your heart and spirit:

>> **Practice meditation.** "Didn't we cover this?" you may wonder. Indeed, but here's the thing: Meditation benefits your heart and spirit too. It helps cultivate self-compassion and spread love — like daily nurturing a positivity garden. There are many types of meditation, such as breath meditation, light meditation, or walking meditation. Explore many of them until you find one that works for you.

>> **Embrace life and people with all their highs, lows, and quirks.** Life's a roller coaster — exciting because of its peaks and valleys. It's the same with people too.

>> **Consider service.** Giving to others is a rewarding way to bond with the world.

>> **Form deep, heartwarming relationships with friends, family, even pets.** They're our anchors, our cheerleaders, our places of comfort. Like your heart, they need tender loving care.

>> **Practice gratitude.** Being grateful for everything that you can — big and small, good and not-so-good. Gratitude is essential for humans because it promotes a positive mental state, deepens relationships, and enhances overall life satisfaction by helping us appreciate our experiences, rather than taking them for granted. It's about appreciating the journey.

>> **Learn from the natural world.** Taking a cue from nature, humans can learn gratitude from its resilience and constant renewal; just as a tree weathers seasons but always blossoms anew, we can appreciate the ebbs and flows of life. In addition, observing the interconnectedness in ecosystems can inspire gratitude for our own networks of support and the vital role each element plays.

>> **Give in to laughter.** We've long known that laughter has physical health benefits. It can also ease tension and brighten your mood. Enjoy the ride; don't take life too seriously.

>> **Tend to your spirit.** If you belong to a faith, prayer is a beautiful way to renew your spirit. So is participating in your religious community. Quiet reflection, acts of service, expressions of art, and mindfulness practices are additional ways you can foster your spiritual growth.

REMEMBER

For those practicing a religion, you'll most likely find that spiritual growth typically forms a key tenet of your faith. But, nurturing your spirit isn't exclusive to religious individuals. Renowned astrophysicist Carl Sagan, who was not known to have religious beliefs, experienced spiritual awe and humility through his study of the vast, complex cosmos.

Focusing on the things you can control

Do you often find yourself stuck in a whirlwind of worries about things you can't control? Well, if you're human, then most likely you do. We all experience that to a certain extent. Even the most mindful, gracious people among us find themselves worrying at times. Here are two ways you can help ease the burden of this worry:

>> **Respond more and react less.** You can't control the actions or attitudes of others, but you can manage your reactions. You're essentially taking charge of your emotional wellbeing, not allowing others' behavior to control you. By focusing on constructive responses rather than succumbing to frustration, you empower yourself, reduce helplessness, and foster a sense of peace.

>> **Worry less.** Easier said than done, right? But here's the thing: Worrying is like sitting on a rocking chair, it gives you something to do, but it doesn't get you anywhere. So try replacing worry with acceptance and action wherever possible.

A constructive response to negative things is not about being positive. It's about coping and providing yourself with a sense of agency that you can work through even the most terrible things. Focus on doing what you can do or control. No one can control the entire world.

Starting with your destination in sight

In whatever situation you find yourself in, think about what outcome you want to achieve. Most often we're so focused on the immediate tasks we have to do that we forget to think about what we want those tasks to lead to. Think about it like setting out on a trip. You wouldn't just board a plane without first knowing where you're going, right? It's the same with all of life's adventures.

A CONSTRUCTIVE RESPONSE VERSUS TOXIC POSITIVITY

Employing a constructive response to negative experiences does not mean that you should employ "toxic positivity" — a response to negative events where people try to force an artificially happy, optimistic state. Doing so can dismiss genuine emotional pain, struggles, and negative feelings. Worse, it can limit your resilience and lead to mental health challenges. So, whenever you see a sign that reads "Good Vibes Only!" remind yourself that your bad vibes (emotions and experiences) *are* allowed.

An example to differentiate constructive response and toxic positivity: Our team member, John, once faced severe abuse from a roommate. Shortly after he was able to move into safer housing, he was beaten and stabbed during a random robbery. Both experiences resulted in trauma. However, he attributes his recovery to a growth mindset and effective habit cultivation.

John says, "The negativity from those experiences was intense. It's absurd to think a positive attitude could counteract abuse or prevent the two liters of blood loss during the stabbing. My constructive response involved recognizing and controlling what I could within these experiences."

He continues, "Amid abuse, I felt helpless. A growth mindset didn't eliminate this, but it helped me identify that feeling, leading to small steps toward seeking external help. Regarding the robbery, a first responder commented that I 'had a positive attitude for someone who just survived attempted murder.' I clarified, 'No, I'm just focusing on what's under my control while letting the doctors and medics do the rest.'"

Develop a clear vision of where you want to end up before you even take the first step. By doing this, you can better align your actions and decisions with your end goal, making the journey more efficient and effective. That's true whether your goal is starting a business, deciding on a career move, or making plans with friends.

Taking a moment to envision your endpoint and planning accordingly makes it easier to address unexpected challenges that pop up along the way.

Prioritizing essentials

Do you often find yourself wondering where your time and energy went at the end of the day? The important thing you wanted to do is not done while you had wasted efforts on things that don't matter much. We've all been there.

To stay on track, you have to prioritize the essentials. Focus on the vital few tasks that align with your goals and say no to the many trivial tasks that don't. Focus on figuring out what really matters in your life and giving those things the lion's share of your time and energy.

As humans, we're champs at overcommitting and at overestimating our capacity. That's why so many of us routinely end up with half-finished weekend projects. Prioritizing helps us gauge what's doable. It's like sorting; you're sifting through your tasks, dealing with immediate and important ones, while spotting the rest that can wait. This approach can be surprisingly soothing. There's joy in ticking off tasks that matter quickly and a certain relief in discovering some things that can wait until tomorrow.

Seeking mutual benefits

Have you ever dealt with a self-centered person who doesn't care about your needs and wants? Sometimes taking the time to understand why they're being unpleasant, rude, or uncooperative can be surprisingly insightful. With a bit of empathy, often the reason is easy to spot, such as stress from being late or stuck in a traffic jam, or receiving unexpected bad news. That may give you enough room to work past their attitude. Other times, you may just need to let them be who they are and instead focus on what you can effectively do.

Not fun, huh? So, should you just reciprocate in the same manner as the self-centered? No, let's be better than that!

Adopt a mindset where you consider others' needs and wants alongside your own.

REMEMBER

The best outcomes happen when everyone feels their needs are met. By practicing this habit, you're not just improving your own situation, but also making others' better. It's the ideal blend of assertiveness and cooperation, which can generate a win-win outcome.

Listening first before speaking

Have you ever heard the expression, "we have two ears and one mouth because we're meant to listen twice as much as we speak"? Though we're pretty sure biologists would disagree, the principle of listening more than speaking still holds up. Following are tips on how to be a better listener:

>> Instead of jumping in with your own opinions and thoughts first, step back and really listen to what the other person is saying.

>> Ask questions to get clarification on things that the other person is saying that don't make sense.

>> Make an effort to understand the other person's viewpoint, their feelings, and their needs before you respond.

>> Reflect on what you heard and check whether they are satisfied with your understanding.

>> With that solid base of understanding, talk about your thoughts and perspectives and how you can collaborate for a better outcome.

REMEMBER

By listening before speaking, you not only show respect and empathy, but you also are likely to have more meaningful and effective conversations. And don't worry; when you have gifted someone with your sincere listening and shown them that you have understood where they are coming from, they are more than likely to be much more open to listen to what you have to say!

Contributing to shared success

Have you ever been part of a project or an activity where everyone's contributions were welcome and things worked out really well? Didn't that feel great? Chances are that something worthwhile got accomplished as well. Being a great collaborator is essential to success, especially in the workplace. Here are some tips to help you be an active contributor:

>> Engage actively! Don't just stand on the sidelines. Jump in, share your ideas, give feedback, and offer assistance in areas you can to propel things forward.

>> If someone is struggling in an area where you excel, be there for them. Humans are a social species, so it is important to help each other out.

>> Remember to celebrate the success of others and to listen for their needs.

You can learn to be an active contributor to every group effort you are a part of. The more you chip in to help your team succeed, the more you get out of it personally and professionally. When you are such a team player, people seek you out to join their teams. It's a sweet deal all around!

REMEMBER

The power of habits cannot be overstated. They play a significant role in your daily life, influences your actions and decisions, and ultimately your wellbeing. The ability to control and change your habits is one of the most powerful skills you can learn.

Achieving Your Goals

In Part 2 of this book, we discuss how executive function — the way our brain gets things done — shows up in various neurodivergent conditions. (Review those chapters if you want to know more.) As we said earlier in this chapter, you can't control others. You can only control your own emotions and actions. In the following sections we focus on how to use your executive functioning to achieve your goals.

Managing yourself effectively

No one knows you like you do. By paying attention to your priorities and your needs, you'll be able to change how you do things so that you can more effectively get stuff done. You're in control. Here are some tips to help you along:

>> **Initiate tasks.** Take the first step to start what needs to get done. Other steps will follow. Don't procrastinate.

>> **Keep track.** Keep tabs on things you need to remember.

>> **Plan.** Break down tasks into manageable parts and determine an efficient sequence.

>> **Manage task and time.** Regularly assess your progress. What have you completed? What remains to be done? How much time do you have? Make any necessary course corrections.

- » **Organize.** Arrange your workspace, physical and virtual, to suit your needs and boost productivity. Don't waste time dealing with clutter.

- » **Cool your emotions.** Check in with what you're feeling. Is it stress, joy, or something else? Acknowledge your feelings, but don't let them dictate your actions.

- » **Shift your approach or focus.** If your approach isn't working, don't get stuck. Adapt your thinking or methods to suit different situations, as they arise.

- » **Monitor yourself.** Keep an eye on what you are doing. Are they helping you with reaching your goal or derailing you?

- » **Monitor your reactions.** Exercise self-control to resist impulsive actions you may regret later. Put away distractions, such as your phone.

Spotting glitches in executive function

Think of executive function as your brain's captain. When it falters, chaos ensues. You may struggle with time management, akin to sailing aimlessly. Planning and organizing tasks may resemble a scattered jigsaw puzzle without a guiding image. Unchecked emotions may feel like a storm-tossed ship with no steerer. Lack of behavioral inhibition can be like a brain without brakes, leading to impulsive actions. Finally, issues with your working memory may make you forgetful, like a sailor losing their course. Challenges indeed, but with the right strategies and support, you can smoothly sail these waters!

Tuning up your executive function

Are you stuck with whatever level of executive function you have? No, not really. As a neurodivergent person, you have a variation in executive functioning that is different than most. That variation is pretty set. But, what you can do is strengthen your own type of executive functioning so that it performs the very best.

It's like having two different car types. No amount of polishing can turn a family sedan into a pickup truck. They're just different cars. But, the owners of each type can still provide their car fuel, keep it tuned up, and wash it so that it performs and looks its very best.

TIP

When it comes to getting your type of executive functioning to perform its very best, here are some tips:

- » Treat your brain as a muscle, building strength through growth mindset and good habits.

- » Break large tasks into smaller, manageable goals.

- » Create commitments to stay on track, such as telling your manager, "I commit to finishing this by Friday."

- » Share goals with a friend for mutual accountability and encouragement.

- » Avoid procrastination; make decisions your future self will appreciate.

- » Maintain a routine for focus and to minimize distractions.

- » Start tasks with clear objectives, such as setting a GPS destination.

- » Ask yourself, is this negative feeling helping me or stopping me? Utilize negative emotions as fuel to propel you rather than an anchor holding you back.

- » Aim high; challenges promote growth. Enjoy the journey even more than reaching the goal.

Building Resilience to Overcome Setbacks

Ever had one of those days when it seems everything goes wrong? Coffee spills, missed buses, grumpy bosses. Your response to these hiccups? That's called *coping* — your personal toolbox for handling life's stressors.

Now, imagine that instead of just a bad day, you're dealing with a major life crisis — a serious illness, loss of a job, a breakup. The bigger the stressor, the harder it can be to cope. But there's something inside you that helps you to not only survive these difficult times, but also thrive afterward. That's your *resilience*. Resilience is your inner bounce-back system. It's your ability to adapt and recover from adversity. Think of your resilience as a piece of elastic. No matter how much you stretch, you always manage to bounce back to your original shape.

REMEMBER

Coping and resilience complement each other. Effective coping strategies enhance resilience, and resilience in turn eases coping. Both are skills honed over time and practice. As you better manage daily stressors, your resilience strengthens. So, even on the toughest days, remember it's part of your resilience-building journey and refining your coping skills. You're stronger than you realize!

Coping with daily stressors

So, how do you cope with daily stressors? Here are a few ways:

>> **Mindfulness** is a multifaceted technique that anchors you to the present, keeping past regrets or future worries at bay. This may be done through yoga, meditation, or simply being present in the moment.

>> **Distraction** can be a powerful ally against stress. Take time out to immerse yourself in activities you enjoy, from painting and running to video gaming.

>> Never underestimate the power of **social support.** Discussing your feelings with friends, family, or a mental health professional can be highly beneficial.

>> **Self-care** should be a priority. Not only does it encompass physical health such as proper sleep and nutrition, but it also includes looking after your mental wellbeing.

REMEMBER

Building coping skills takes time and practice and figuring out what works best for you. But once you've got them, they'll help you navigate life's challenges.

Building your long-term resilience

Fostering resilience is a practice, not a one-and-done solution. So, what types of things can you practice to build your resilience? Here are a few:

>> **Prioritize relationships.** Foster connections with family and friends for support during tough times.

>> **Embrace acceptance.** Understand that life comes with challenges instead of expecting constant smooth sailing.

>> **Maintain perspective.** In difficult times, remember that most problems are temporary and surmountable.

>> **Engage in enjoyable activities.** Self-care boosts resilience and provides strength to help others.

>> **(Again) Prioritize self-care.** Regular exercise, nutritious food, adequate sleep, and meditation significantly enhance resilience.

REMEMBER

Fostering inner strength or resilience isn't about avoiding difficulties; it's about learning how to cope with them better. It's an ongoing journey, not a one-time deal.

4

Thriving as a Neurodivergent Person

Explore what it means to thrive as a neurodivergent person and understand the conditions that enable you to feel comfortable with who you are, use what makes you unique to your advantage, and genuinely enjoy your life.

Find out how you or your child can thrive in an educational setting from kindergarten through higher education.

Discover strategies for finding work and nurturing your professional development, as well as strategies for how to be effective at work and become a great employee.

Look at how neurodivergent individuals can develop and understand interpersonal relationships, including how to navigate the dating world, find community, and connect with friends.

Chapter **13**

Discovering What It Means to Thrive

For the past two years, Alex has watched over a houseplant that sits in her bedroom window. She took great care to position it in a shaded corner where it wouldn't be damaged by the harsh light of the midday sun. Alex waters it, trims it, checks it for bugs, and occasionally gives it plant food. And in response to all these efforts, the plant survives and gets by. It's not the healthiest plant, but Alex thinks "it'll do."

One afternoon, a friend mentions to Alex that this particular type of plant isn't a houseplant at all. Perhaps its label had gotten mixed up at the shop. This type of plant actually thrives best when planted outdoors, soaking up the heat, and basking in bright, direct light. So, Alex moves the plant to her garden — having learned about its particular needs. As days pass, and now under the right conditions and care, the plant grows slowly at first but then really takes off. When handled as a houseplant, its leaves were brittle, its stems weak. But now outside, it pops up leaf after leaf and suddenly blooms. It's still the same plant, but where it once merely existed, it's now vibrant. It doesn't just live; it *thrives*.

Thriving is an important aspect of life — for everyone. But learning how to thrive is doubly important if you're neurodivergent. Though sitting on a windowsill works great for most, you may need a different approach.

This chapter is all about understanding the conditions that enable you to thrive as a neurodivergent person. It's not about changing who you are but understanding how to come into your own. So, roll up your sleeves and start to cultivate the things that need to be done. It's about turning a life that's merely surviving into one that's richly thriving. So, let's get ready to grow!

The Keys to Thriving: Coping and Resilience

Thriving as a neurodivergent person means more than just getting by. It's about feeling comfortable with who you are, using what makes you unique to your advantage, and genuinely enjoying your life. Thriving builds upon two key things: coping and resilience. *Coping* is how you handle everyday challenges, such as seeking a quiet spot when things get to be too much. *Resilience* is about bouncing back from the hard stuff, learning from it, and keeping on going (there's more on how to do this in Chapter 12).

Thriving is also about finding the things that enable you to feel at ease and really shine. It's about knowing what you need, such as extra time for tasks or a peaceful place to focus, and not being afraid to ask for it. It includes having the confidence to be your true self and recognizing that your unique way of thinking is normal and needed. And thriving means defining happiness and success on your own terms.

Understanding What You Need to Thrive

To thrive as a neurodivergent person, you need to do a few key things right up front. We dig into them here.

Understanding yourself

The first step toward thriving as a neurodivergent person is understanding yourself. This includes learning about how your brain works and how it differs from others. It also means being comfortable with that. It's noticing what makes you happy and what stresses you out and getting to know your challenges and figuring out your strengths.

By understanding these things, you can start to make choices and changes that help you thrive. Such changes may be finding a quiet spot when things get too noisy, taking breaks when you need them, setting expectations with coworkers or friends, or doing activities that you really enjoy. The key is to know yourself and use that knowledge to make your life better.

REMEMBER

As a neurodivergent person, it's important to remember that you live in a world that isn't always set up for you. This may mean you haven't really had the chance to notice or explore what works best for you. Understanding your own way of doing things is a crucial step in beginning to thrive.

Andy Burgess is a documentary filmmaker in the United Kingdom. As a dyslexic thinker, he long struggled to learn in ways that worked best for him. The school system in his hometown just wasn't set up that way. It was only in college when he started playing in bands that he discovered ways of learning that focused on his strengths.

"From that I learned business skills and creative skills and marketing skills and how to promote our band and essentially make it a business," Burgess said in a 2021 YouTube video titled "The Advantages of Having Dyslexia." "That's where I really started to find my passions and what was inspiring me to stay in the creative field. I always had told myself I was going to do something creative, and I wasn't going to let anything else stop me."

Of course, understanding yourself isn't just about understanding your strengths. Recognizing and accepting your challenges helps just as well.

"One thing I learned about myself is I'm not good at dealing with details," said Catherine Bigonnesse in the 2017 *Huffington Post* article, "Five Ways I Learned to Thrive as An Adult Living with ADHD." Bigonnesse, a writer in Canada with ADHD, encourages others to embrace both their challenges and strengths in order to find success. "I learned to structure my workdays in alternating boring tasks and interesting ones (at least from my brain's perspective). That way, I can keep my attention and focus throughout the day and be productive."

REMEMBER

Don't be afraid of understanding what challenges you. Those things are what help you find your limits and set the parameters that create the best conditions for you to utilize your strengths.

Understanding the neurotypical world

If you haven't noticed, our world is largely designed around the neurotypical mind. This shows up in all sorts of ways — such as the ways we learn, work, and even how we chat and hang out with friends (for more on all of those ways, see Chapter 4). Although no one intentionally set out to do this, understanding this fact is an important part of learning how to thrive.

Keep in mind that you're not broken; you're just potted on a windowsill of neuro-typical expectations according to instructions written without you in mind. That *windowsill* may be great for others and not so much for you.

"At the end of the day, anyone who is regarded as 'different' is living in a world that is not really accommodating of them or built for them," said Hari Srinivasan, a minimally speaking autistic PhD student in neuroscience, in a 2020 article published on Medium for the Project Illuminate Initiative. "So, it's a constant state of trying to fit in, which is very stressful and can lead to burnout."

Understanding your normalcy

There's nothing special about your neurodivergence. No, really. There isn't. You're just as boring and ordinary as every other person on our planet. And guess what? That's a great thing!

Your neurodivergent traits aren't special (hooray!)

Your neurodivergent strengths are not superpowers, nor are the challenges you face epic tragedies. They're simply what they are. Every person has traits that can prove an advantage, and all of us have challenges we sometimes wish we could change. This is called *being human*.

Now, your challenges may be extremely tough. We get that! But here's the thing: Life is challenging for all sorts of different people in all sorts of different ways. Our *flavor* of challenges doesn't make us special, nor does our mixture of strengths. Strengths and challenges are normal parts of life.

As a neurodivergent person, you're just as normal as everyone else. There's no need to prove that fact. It's simply true. Should we say it again? Maybe we should pop it out as a stand-alone sentence so that it really sinks in:

As a neurodivergent person, you're just as normal as everyone else.

There! You see? There's nothing special or spectacular about your neurodivergence. Just as there's nothing special about the color of your eyes, how many hairs you have on your head, whether you still have your wisdom teeth or not, or how tall you are. Can you appreciate these things about you? Of course! But it doesn't change the fact that these characteristics are all ordinary things. There's nothing particularly special about you being born the person that you are.

You can still be spectacular

That doesn't mean you can't be spectacular! Your neurodivergent traits may be ordinary, but what enables you to stand out is how you use your unique perspectives and skills. Your neurodivergence gives you a different way of seeing and interacting with the world. Be thankful for that! Lean into those traits. Maybe you're really good at spotting patterns, thinking outside the box, or coming up with creative solutions others may not see.

When you embrace your neurodivergence as normal, you gain confidence in the traits that enable you to be you. Great at focusing on details? Apply it to tasks that require precision and accuracy. Able to see things from a different angle? Incorporate that into situations that require problems to be solved. Have difficulty keeping track of time? Find the mechanisms that can accommodate that challenge. The key is embracing your neurodivergence as a normal part of life.

REMEMBER

Everyone has strengths and challenges, whether neurodivergent or not. Accepting your own enables you to dazzle yourself and others with your unique talents and skills. It's not about being *special* but about understanding your challenges and leaning into your strengths. Accept who you are and use it to thrive.

A hard lesson to learn

We hope you're beginning to get that there is power in understanding yourself as normal. But that doesn't mean it's an easy lesson to learn. It's something you have to keep working on. There may be times when you doubt yourself, and that's okay.

You may have had a lifetime of feeling as though you were an outsider. Others may have said that to your face. When this happens, it's tough not to start believing it yourself. Even if you don't consciously think you're *less than* others, a part of you may still feel that way somewhere deep down.

"It took many years of vomiting up all the filth I'd been taught about myself, and half-believed, before I was able to walk on the earth as though I had a right to be here," said James Baldwin, writing in his 1960 essay in *Mademoiselle* magazine, "They Can't Turn Back." Baldwin was one of the greatest thinkers of the 20th century, shaping how generations of people understood their identity and their place in this world — and it still took him time. It's okay if it takes you time too.

Relax: No need to prove yourself

Too often we feel the need to flash our neurodivergent strengths to prove we're just as normal as everyone else. It's as though we need to go out of our way to say "Hey! I deserve to be here too." Relax. There's no need to do that.

REMEMBER

Realizing that you're normal helps you find your true strength. You'll see that it's not your neurodivergent traits stopping you from being understood, accepted, and included, it's actually other people's attitudes and unnecessary barriers. When you understand that your neurodivergence isn't the problem, you start looking for what really needs to be fixed.

This mindset takes away the pressure to prove that you're *special* and lets you be your true self. It's empowering to know that, like everyone else, you have your own strengths and challenges and that's okay. If others don't see it that way, well that's on them. Brush your shoulders off and move on.

Accepting your right to thrive

Every human being deserves to thrive. As a neurotypical person, you shouldn't have to fit into the neurotypical world or ignore your own needs to do so. Just like anyone else, you deserve to live your life in ways that allow you to be happy and fulfilled. Accepting your right to thrive begins with truly believing that you deserve it. This is not just a pleasant idea, it's a fact. When you believe this, you can better understand and fulfill your own needs and wants. It's about claiming your space in the world, feeling that you belong, asking for what you need, and setting up the right conditions for your growth and progress.

REMEMBER

Thriving is more like exercising than simply a way of thinking. It's something you actively do, not just something you feel. As with many things in life, it takes effort. Think of thriving as a daily practice where you use your strengths, face your challenges, and make choices that lead to a satisfying life.

Tapping Into Support Systems

Think back to Alex's plant we mentioned at the beginning of this chapter now vividly living in the garden. It just doesn't thrive on its own; it needs support to grow. To be its gorgeous, luscious self, Alex's plant requires sun, rain, sometimes extra watering, and depending on the season, it may need a scoop or two of plant food. And although it no longer lives on her bedroom windowsill, the plant in Alex's garden still needs her from time to time.

You're just like that plant in that you need things beyond yourself to help you grow and thrive. Actually, we're *all* just like that plant — no one person grows on their own. All of us need support, resources, and folks who understand and accept us. In the sections that follow, we talk a bit about how you can find and make use of a wide range of support systems to help you grow.

Using accommodations

Every single human uses accommodations (for more on that, turn to Chapter 2). As a neurodivergent person, you need accommodations too. Some of these accommodations may be related to your neurodivergence, while others are related to other aspects of your life — such as cars and public transportation, online shopping, remote-work options, meal-delivery services, and workout gyms.

In simple terms, an *accommodation* is any product, adjustment, or service that makes a task or situation easier and more comfortable for someone to handle in their daily life. We all use them. Cars and public transportation make getting to places easier and more comfortable, online shopping makes acquiring things we need easier, remote-work options enable us to be more productive, meal-delivery services may help save us time and effort, and workout gyms help keep us fit without running through the forest or doing hard labor all day.

REMEMBER

As a neurodivergent person, using an accommodation as it relates to your neurodivergence is no different than using an accommodation for any other aspect of your life. Accommodations are normal, and using accommodations to assist your neurodivergent traits is utterly ordinary too.

Think about what you need as a neurodivergent person to do the things that enable you to thrive in life. You may notice some of them right away, but figuring out others may take a little time. Here are some accommodations neurodivergent people often use:

>> **Quiet workspaces:** Having a calm and low-noise area to work or study.

>> **Flexible scheduling:** Adjusting work or school hours to better suit personal needs.

>> **Clear instructions:** Receiving directions in a straightforward, step-by-step manner.

>> **Extended time:** Allowing extra time for tasks or tests.

>> **Visual aids:** Using charts, pictures, or diagrams to support understanding.

>> **Breaks:** Taking short, frequent breaks to manage sensory overload or stress.

>> **Written communication:** Preferring email or texts over phone calls.

>> **Routine and structure:** Having a predictable schedule or environment.

>> **Sensory tools:** Using items such as noise-canceling headphones or stress balls.

Accommodations can also be used in social situations, such as when interacting with your family or hanging out with friends. This may include the need for flexibility in plans or recognizing that sudden changes in environment or schedule can be overwhelming. Some may appreciate having alone time to recharge, especially after engaging in lengthy social interactions. At gatherings such as parties, having a quiet place to retreat to can be crucial for managing sensory overload or simply taking a break from the hustle and bustle.

Clear and direct communication can also be a helpful accommodation, as it reduces the chances of misunderstandings. So is the use of closed captioning while watching a movie or television, which more and more people are finding to be a helpful accommodation. These accommodations help you enjoy social and family situations while also respecting your personal limits and maintaining the balance you need to thrive.

Utilizing resources

Think of resources as tools, services, or people that offer support, information, or skills to help you navigate life more smoothly and thrive as a neurodivergent person. Resources also include support groups where you can share experiences, tips given by others who understand your perspective, educational materials such as books or online courses that can teach you helpful strategies, public services offered by your local government, and help from professionals such as therapists or coaches who specialize in providing services from a neurodiversity-affirming frame.

TIP

Finding these resources can start with a simple online search, asking for recommendations from others in the neurodivergent community, or reaching out to organizations dedicated to neurodiversity support (see the appendix for some great starting points. To quote the Barefoot Contessa, Ina Garten: "How easy is that?").

REMEMBER

Once you've identified potential resources, securing these resources may involve some legwork, such as making calls or filling out forms. But the payoff from your effort can be huge. Resources can help you address challenges that come your way more easily or provide you with the support needed to really tap into your strengths.

Mapping your support network

If you're neurodivergent, you may sometimes feel that you're all alone. A lot of us grow up feeling that way, not realizing there are role models out there who've been through the same things (more on that in Chapter 3). And too often, we're told we're *broken* or need *fixing*, which isn't true at all. We're just different, and that's okay.

If this is you, it can be tough to reach out for support. Sometimes asking for help may feel as though you're admitting something's wrong, or you may worry about being misunderstood or judged. But remember, reaching out is actually a strong and healthy thing to do. And if you want to thrive, then mapping out and building your support network is key.

"Independence is almost like this myth we've been sold," said Haley Moss, a Miami-based attorney and author of *The Young Autistic Adult's Independence Handbook,* on the June 13, 2002 episode of the *Beyond 6 Seconds* podcast. "I thought independence meant I had to do everything myself. I generally thought it meant no assistance. I thought it meant that I wouldn't have to ask my parents for things." Moss, who is autistic herself, eventually realized that even the most *independent* adults rely on support networks.

One of the beautiful things about support networks is that support flows both ways — sometimes you're the one getting help, and at other times, you're the one offering support. Moss continues: "In some places, and some pieces of my friend circle, I'm now the responsible adult friend, which I think is ridiculous, because I don't think I know everything." None of us know everything, nor can we do everything ourselves. That's why we need networks of support!

So, think about the people and places where you feel safe and understood. Maybe it's a friend who always has your back, a teacher who understands your learning differences, or a counselor who offers great advice. Online communities can be a big help too, especially forums or groups where people share similar experiences.

TIP

Start by considering what kind of support you need and who may be able to provide it. As you map your support system, here are types of support you may find helpful:

>> **Emotional support:** This support may come from a spouse, a family member, or any number of friends. It may also include resources such as neurodiversity-affirming therapists. Really, emotional support may come from anyone who unconditionally accepts you for who you are (and don't forget about the unconditional acceptance and emotional support you often find in your pets!).

>> **Practical support:** This support comes from such places as student services offices, human resources departments, advocacy organizations, and government offices and programs.

>> **Social support:** This support comes in a lot of forms from gathering with friends, to sports leagues, trivia teams, or book clubs. These social gatherings don't have to be about your neurodivergence (although there are lots of neurodivergent social groups you can join). The point is to connect with activities and people who resonate with you.

>> **Informational support:** This support includes those items that provide the information you need to navigate the world and better understand yourself, such as books, websites, or workshops about neurodiversity.

Community: Finding People to Nurture You

Humans, by nature, crave connection and community. It's like how plants thrive best when they're around other plants that share their environment and needs. For neurodivergent people, finding a community of others who understand you is incredibly important.

"Finally, I could stop hurting myself, blaming myself for not fitting in," said Barb Cook, recalling in a 2021 article in *Spectrum Women* magazine what it felt like to receive her autism diagnosis in 2009. "I didn't need to fit in; I just needed to find my people." And that she did. Cook, who is also dyslexic and has ADHD, is the founder of *Spectrum Women* magazine and the director of Australia's Neurodiversity Hub.

REMEMBER

When neurodivergent individuals connect with others who share similar experiences, they find support, friendship, and a sense of belonging. These connections can help you grow, just as the right soil and sunlight can help a plant thrive. Being part of a community that truly understands can be a game-changer in thriving and flourishing in life.

Online and in person, there are endless networks of neurodivergent people who share experiences similar to you. And there are lots of ways to connect (check out the appendix for more). As you continue your growth, make sure you're connecting with others like you.

Cultivating Self-Advocacy

To truly thrive, it's essential to embrace self-advocacy. This means learning to recognize your own needs, values, and rights, and then actively speaking up for them. By doing so, you become more confident and capable of steering your life in the direction that suits you best.

Understanding self-advocacy

At its core, *self-advocacy* means understanding your needs and how you work best — whether that's in school, at work, or in social situations with friends or family members. It also means recognizing that your needs are normal, and then speaking up for them.

Imagine that you're at a restaurant and the waiter gets your order wrong. You requested a bowl of rich and creamy tomato soup and instead you were served a cold lump of cottage cheese. Self-advocacy is politely getting the waiter's attention and saying, "Excuse me, this isn't what I ordered. May I have what I asked for?" It's about communicating your needs clearly and effectively.

Self-advocacy can be as simple as stating, "I work best when _____" or "I learn best when ___" or "If you hear me speaking too fast, please ask me to slow down" or "I would like more clarity on this topic." It can also take the form of phrases such as "I need a bit more time to get ready if you want me to meet you for lunch," or "This coffee shop is a bit too loud for me to hear you," or "I just need some alone time tonight."

REMEMBER

Self-advocacy helps you clear up misconceptions before they occur, maintain your boundaries, set clear expectations, and identify the parameters you need to grow. And advocating for yourself allows you the opportunity to learn and discuss the needs of those you care about as well. No one knows you better than you do. You are your own best spokesperson.

Understanding neurodivergent self-advocacy

Think back to the example we mentioned in the previous section, where a waiter brings you a dish you didn't order. As a neurodivergent person, you may find that situations like this occur frequently. The world doesn't always understand or support your needs. You'll need to speak up to make your needs known. This may involve explaining how you process information or what environments help you work best. It may mean requesting specific support at your job or school, such as a quiet space or extra time for tasks. It may also involve setting clear expectations with your family and friends.

Self-advocacy may also mean setting boundaries. These are the limits and rules you create about how you want to be treated. Boundaries help you respectfully tell others what you're okay with and make sure your needs and preferences are respected. For example, you can set boundaries by letting people know when you need alone time, specifying how you like to communicate, or asking for space when you feel overwhelmed.

It's important to note that setting boundaries can sometimes be tough. This is especially true for neurodivergent people conditioned to always do things the neurotypical way. On the social networking site X, writer Callum Stephen Howes, who is both autistic and ADHD, explains: "We're so used to being uncomfortable and having to give so much to every situation that we might not know where to draw the line."

REMEMBER

Boundaries may be limitations, but they create opportunities for growth. Think of them as similar to a garden trellis that supports and guides a plant. Without it, the plant may get stepped on or overshadowed by others. With boundaries, you establish structures that allow you to continue to grow.

Here's a cool thing: Advocating for yourself means not just asking for what you need, but also being proud of who you are. It's about valuing your own experience and using it to find your way in the world. It's loving yourself and confidently saying, "Hey! I am what I am."

"I can't expect everyone to see and accept me as I am," said Emily Chen, an educator and classical musician with ADHD, in a 2021 article in *ADDitude* magazine. "But I can do my part to support myself by building my sense of self-worth and confidence in my voice bit by bit, one day at a time."

TIP

Self-advocacy doesn't just benefit you — it also helps the people around you. When you speak up for your needs and explain how you see things, it gives your friends, family, and others a better understanding of who you are. They learn how to support you better and what makes you comfortable. This doesn't just make things easier for you, it also makes your relationships stronger. By advocating for yourself, you're helping others to learn and grow too.

Chapter **14**

Thriving in Educational Settings

Thriving in educational settings as a neurodivergent individual is much more complex than just doing well academically. Your educational journey also involves personal growth, emotional well-being, social and communication development, learning essential life skills, maintaining relationships with peers and instructors, and learning to advocate for yourself. Education is about growing as a person, not just growing your knowledge.

This chapter lays out how a neurodivergent person can thrive in pursing education. Whether you are a K-12 student, a student in high school or higher education, or pursuing a lifelong education as a neurodivergent individual, this chapter outlines tools and strategies that can help you better understand yourself, grow as a student, and connect with others in the educational setting in meaningful ways. For parents raising neurodivergent children, this chapter can help you navigate your child's educational path and secure the resources you both may need.

Understanding Yourself or Your Child as Neurodivergent

If you want to thrive as a neurodivergent person, the first step you need to take is recognizing yourself as neurodivergent. It's at that point of understanding that you begin to lay out your own personalized blueprint for how to best navigate life. Recognizing oneself as neurodivergent isn't a single aha moment. It generally begins by questioning yourself over time. As the parent of a neurodivergent child, you may also go through this process of questioning before coming to the understanding that your child is neurodivergent.

REMEMBER

Many, *many*, neurodivergent adults move through life today unaware that they are neurodivergent, or who are in various stages of questioning themselves, trying to figure out what exactly is that unnamed "different thing" about them that they feel, and searching for how they fit into our larger world. If you are neurodivergent and enrolled in education, now is a great time to begin figuring yourself out and how you can be the best you can be in this complex world.

In the following sections, we walk through the educational journey from elementary school to university to see how understanding oneself or your child as neurodivergent can greatly influence their educational path, enabling tailored approaches and support that cater to their unique learning needs.

Enhancing the Elementary School Journey

In elementary school, when children are still piecing together their self-concept, grasping their neurodivergence can significantly boost their confidence. They learn that their brains just operate differently, not wrongly. This revelation can drive them to better engage with their unique educational needs, fostering a more inclusive learning experience right from the start.

WARNING

It can be tempting for parents to hide their child's neurodivergence from their child. When this happens, it's mostly done out of love and in an attempt to protect their child from harmful things such as bullying, stereotypes, or self-doubt. However, this can hinder the child from understanding themselves, developing coping skills, and limit their opportunities to grow.

TIP

What's the alternative? *Talk about it!* Break down discussing the child's neurodivergence into kid-friendly language, emphasizing that it's not a roadblock but a unique way of thinking and seeing the world.

Our ultimate aim is to equip neurodivergent individuals with independence and control over their own lives. However, the truth is, when you're young, your caregivers largely shape your world. For neurodivergent children in elementary school, ensuring they get the resources necessary for their educational success falls on someone else's shoulders.

TIP

Here are some valuable tips for parents and caregivers striving to enhance their neurodivergent child's elementary school journey:

>> **Understand your child's needs.** Work to understand your child and their needs so that you can identify and advocate for accommodations that can help them thrive.

>> **Foster self-awareness.** Help your child understand their neurodivergence in an age-appropriate way, emphasizing their unique strengths.

>> **Teach self-advocacy.** As your child grows, gradually teach them to advocate for their own needs.

>> **Encourage their interests.** Support and encourage your child's passions. These can be great avenues for learning and social interaction.

>> **Create a consistent routine.** Neurodivergent children often thrive with predictable schedules. Establish a regular routine both at home and school.

>> **Introduce them to other neurodivergent children.** Neurodivergent children often communicate and socialize in ways easily understood by others with similar experiences.

>> **Engage with school.** Actively participate in school activities and meetings, making your presence known. Regularly check-in with teachers and maintain a positive, solution-focused approach. An individualized education plan (IEP) is tailored to meet the unique educational needs of the child. It outlines specific educational goals, services, supports, and accommodations necessary to ensure a child's academic success. Explore the need for an IEP for your child and if applicable, actively participate in the development of your child's IEP with their education team.

>> **Seek support.** Connect with neurodiversity advocacy groups and with other parents for mutual understanding and support.

REMEMBER

Many neurodivergent children experience trauma from growing up in a world that is not designed for them. Unfortunately, most schools are not trained in how to recognize this, nor are parents provided necessary information to address it. If your child has experienced trauma, the impact doesn't have to be permanent, and practical steps can help them heal and grow. (For more on this, turn to Chapter 12.)

Advocating for Yourself in High School

As a neurodivergent high schooler, it's vital to understand that your unique needs not only shape the way you learn but also contribute to the unique lens through which you view the world. This period of your life brings new challenges and opportunities, but remember, this is your story to write.

WARNING

When navigating high school as a neurodivergent individual, you may find that the relationship with your parents experiences a certain tension. This can be a natural part of adolescence, as you're striving for independence while they may still see you as their little one they want to protect. Remember, they're learning too, trying their best to balance between protecting you and letting you grow. While it can be challenging, try to approach these tensions with understanding. Open, empathetic communication can help bridge this gap and foster mutual understanding. *And remember* — if you decide to have children in the future, you'll likely understand this tension from your parents' side!

TIP

As you move through high school, here are some tips that can help you thrive:

>> **Understand your needs.** Identify your unique learning needs and social preferences.

>> **Advocate for yourself.** Speak up about your needs with teachers and counselors.

>> **Engage in activities.** Join clubs or teams that align with your interests and strengths.

>> **Be patient with parents.** Recognize the love and concern behind their actions.

>> **Build a support network.** Surround yourself with positive, understanding friends and mentors. Regularly discuss challenges and wins with people you trust.

>> **Plan ahead.** Start thinking about your plans after high school, such as work, college, or vocational education. Talk with your parents and school counselors about this.

>> **Celebrate your neurodivergence.** Embrace your unique perspective as a strength.

>> **Trust your journey.** High school is a time of self-discovery. Embrace your path, even if it looks different from others'.

>> **Foster resilience.** Accept setbacks as part of the journey, learning and growing from each experience.

Navigating Post-Secondary Education

Whether you're attending a university, community college, or a vocational school, your post-secondary education can be an exciting ride filled with newfound independence and self-reliance. This means taking care of yourself, from meals to errands to laundry, which can be a fun but challenging part of being an adult.

Academically, you'll dive deep into your chosen subjects. Expect lots of self-study and long hours of learning that at times can feel exhausting. However, there also can be an excitement and joy that comes from learning new things, discovering new concepts, and acquiring new skills.

The social aspect is vibrant too. Often, this is the time in life when people meet others with vastly different life experiences than their own. You'll meet people from diverse backgrounds in classes, clubs, dorms, or just hanging out. It's not just your classes that you'll learn from. You'll learn new perspectives, approaches, insights from these classmates too. All of these social connections contribute to your own growth as a person and broaden how you see and interact with the world. (Check out Chapter 16 if you feel hesitant or anxious about social interactions.)

Personal development is the term used to describe the process of improving oneself. This includes acquiring new knowledge, gaining new skills, learning more about yourself, deepening your values, and strengthening your own character. While you won't be graded on it, personal development forms a huge part of your post-secondary education. From learning how to manage group projects to finding joy in the discovery of new ideas, your post-secondary experience is an amazing time to grow.

REMEMBER

Education after high school is a time to stretch your independence. That can be rewarding in so many ways, but it also means that the responsibilities of managing your own life and actions are shifting more heavily to you. As children, our parents or caregivers were largely responsible for us. At this point in life, its often necessary to figure out how to live and thrive on your own (even if you're still living at home). Stay open and flexible as you do this. You'll encounter challenges, setbacks, and at times some awful things. This is all part of life.

Each of your classmates has their own strengths and challenges. If you're working your way through post-secondary education as a neurodivergent person, your challenges may sometimes relate to that. That's not something to fear; just to figure out. You're not alone. We guarantee you that there are other students with similar experiences to yours. It may be helpful to seek them out and connect. There are resources and practices that can help you — check out the appendix for some ideas.

No matter what type of education you're pursuing, here are some tips that can help you out:

>> **Leverage campus resources.** Use academic support centers, tutoring services, and mental health counseling available on campus.

>> **Advocate for yourself.** Speak up or write down your needs in the classroom and discuss with professors. Request accommodations if necessary.

>> **Discover effective learning strategies.** Find out how you learn best, and implement these strategies in your study habits.

>> **Maintain a balanced lifestyle.** Ensure a balance between academic work-load, social activities, self-care, and rest.

>> **Build a support network.** Form relationships with understanding peers, mentors, and educators who can offer support and guidance.

>> **Plan and organize.** Use tools such as planners or digital calendars to manage your time effectively.

>> **Seek career advice.** Utilize career services to explore future job prospects and gain practical advice.

>> **Maintain open communication.** Regularly express your feelings and challenges to trusted individuals.

>> **Prioritize mental health.** Consider regular check-ins with a mental health professional, especially during stressful periods.

A WORD TO PARENTS

It's important that you recognize that as your child progresses through higher education, the strategies they've learned to foster independence, celebrate small victories, and practice empathy and compassion toward their own differences can greatly enhance their life. These skills empower them to be authentically themselves, successfully navigate various educational environments, and bolster their personal development and self-acceptance.

In the whirlwind of parenting, it's not uncommon for you to forget to nurture your own identities and needs. Your lives become so entwined with your child that your self-care and personal growth often take a backseat. As parents of a neurodivergent child, you may face additional responsibilities and stressors that necessitate even more time and energy. However, self-care and personal development aren't luxuries for you; they're essential. (We discuss this further in Chapter 19.)

Chapter **15**

Thriving in the Workplace

t used to be that many of the skills you needed to thrive in the workplace were formally taught in school. For example, classes in the industrial arts — also known as shop class — instructed students in using tools and managing projects. Home economics classes taught such skills as accounting, budgeting, and cooking. Unfortunately, these types of classes are offered less and less in school today. One negative result of this has been to shift the teaching of these everyday skills away from the school and onto the student, who is expected to just pick up these skills on their own.

If you're a neurotypical person, you may be able to pick up these skills relatively easily and use them in the workplace. If you're neurodivergent? Well, it can become an almost impossible task to learn such skills when you're already having to spend time and energy learning how to get around all the other barriers a neurotypically designed society places in front of you.

In this chapter, we offer tips, techniques, and best practices that can help you thrive as a neurodivergent employee in the workplace. We include strategies for not only finding work and nurturing your professional development, but also strategies for how to be effective at work and become a great employee. Many of the skills we discuss in this chapter aren't specific to neurodiversity, but rather skills that are helpful for everyone who wants to thrive at work.

Formulating a Strategy for Employment

Transitions — especially when it comes to making a change in your employment status — often carry a feeling of risk. They upend the familiar and thrust you into the realm of the unknown. That scary feeling associated with risk is often amplified if you are neurodivergent, as you may already be struggling with having to deal with a world designed for neurotypicals. There is a sense of safety in the familiar, the predictable patterns that offer comfort amidst life's loud noise. But, this caution can deter you from job-seeking, make you linger in roles you've long outgrown, or make you shy away from pursuing bigger responsibilities and better pay.

There's an undeniable charm in the human tendency to just wing it. This willingness to leap before looking can be thrilling, much like playing the slot machines at a casino — there's always a chance of winning big, but also a considerable risk of losing it all. To those adventurous go-with-the-flow types, we salute you. However, when dealing with life's realities, such as applying for a job, a planned approach often turns out to be more effective. When you enter a situation with preparation, you're not leaving it all to chance; instead, you're actively tipping the scales in your favor, boosting your odds in the pursuit of success.

REMEMBER

Career development calls for striking a balance between just winging it and safety. Whether you're pursuing your first job or wanting new challenges, making a plan can empower you to explore fresh possibilities without feeling excessively put at risk.

So, how do you both leap *and* look? By taking a conscious approach to advancing your career and developing an employment strategy. Here are some ways developing a strategy can put you at an advantage:

- » **You have a sense of direction.** Without clear direction, it is easy to get overwhelmed. A plan induces calm.

- » **You know where to sharpen your skills.** With a goal, you develop your skills toward where you want to go.

- » **You are able to job search efficiently.** You waste less time by focusing on what is most relevant.

- » **It is easier to have job satisfaction.** Aligning your career with your passions and strengths leads to jobs you'll love

- » **You build resilience.** Clear goals mean job rejections or setbacks are just bumps, not dead ends.

Having a personal strategy for work can give you the direction, focus, and confidence you need to build a career that's fulfilling and rewarding.

Assessing your strengths

As you begin to build your employment strategy, it is important to examine your skills and strengths. Start by asking yourself the following questions:

>> **What do I do well?** Reflect on the things you are good at. It can be anything from problem-solving to teamwork to communicating big ideas clearly to technical skills.

>> **What fuels my energy and sparks my deep passion?** Look for those things that light a fire in you and make work feel less like work, such as connecting with people, crunching numbers, or creating content.

>> **What makes me stand out?** Figure out what makes you different. Try to pinpoint a few key qualities that make you stand out from the crowd.

>> **What accomplishments am I the most proud of?** Reflect on the accomplishments that make you the most proud, whether big projects or small acts, such as consistently treating others with respect. Thinking about these things can help identify which aspects of work are the most meaningful to you.

>> **How do I handle obstacles?** Honestly assessing how you handle stress and setbacks can strengthen how you respond to challenges ahead.

>> **What are my strengths?** Get comfortable with articulating and discussing your talents and skills. If you are not sure of your strengths, explore skills assessment tools such as https://high5test.com to identify your top strengths.

Asking yourself these questions can help you identify your core values, decide which opportunities you may want pursue, navigate workplace complexities, and become more comfortable with discussing your achievements and all the things you do best.

TIP

Check whether your current resume, LinkedIn profile, or personal website accurately highlights your strengths and values. If they don't, give them a brush up. (See the section "Developing a stellar profile on LinkedIn" later in this chapter for more on how to do this.)

Setting goals (and acting on them)

One of the most remarkable stories in human history happened around 1500 BCE. It was then that a small group of adventurous families near present-day New Guinea walked toward the ocean, huddled themselves into double-hulled canoes, and pushed off from the shore. These adventurers, and generations of their children, would eventually come to settle thousands of undiscovered islands stretched out over ten million square miles.

This amazing group had no guaranteed knowledge that they would find new islands when they ventured out into the vast Pacific Ocean. Their voyages were a combination of intentional exploration, luck, and accumulated knowledge. Sailing out, the ancestors of today's Polynesians charted their course by the rotation of the stars, noting currents and the way in which the wind pushed clouds across the horizon, and by observing the movements of migrating birds above and fish below.

Their journeys, much like our own career paths, were filled with both uncertainty and the allure of potential. And just as the Polynesians found their way by observing the world around them, we too can chart our professional course with intention, drawing inspiration from the lessons they left behind. Setting clear goals and timelines is our modern-day constellation, directing us through a broad sea of career opportunities.

Start by establishing SMART goals

When setting goals, many of our neurodivergent students find the SMART method helpful, especially as they seek employment. SMART stands for Specific, Measurable, Achievable, Relevant, and Time-bound. This approach ensures that your goals are both clear and achievable, helping you to not get lost at sea. Following are four examples of different individuals using SMART goals to grow their careers.

Mia is a budding entrepreneur with a dream of starting her own business. Armed with ambition, Mia formulates a SMART goal. Over the next year, she's focused on developing a comprehensive business plan, identifying target markets, and securing a business loan (Specific, Measurable). Her goal is to launch her business, a local coffee shop in a neighborhood that has long desired one (Achievable, Relevant). Mia wants to open her shop within the next 18 months (Time-bound). Through this SMART strategy, Mia moves closer to making her dream a reality.

Now, imagine Sophie, who works in human resources but has a passion for graphic design. Sophie has set a goal of completing an online course in advanced graphic design techniques within the next six months (Specific, Measurable), giving her the skills required for an upcoming job opportunity in her company's graphics department (**Achievable**, **Relevant**), which her company has stated that it will be hiring three people for eight months from now (**Time-bound**). By establishing a SMART goal, Sophie has identified a path to help her transition to a design-centered role.

Next up: Todd, who works part-time cleaning a local paper company's offices three nights a week. His employer has stated that a full-time janitorial role is possible if Todd shows dedication to his work. So, Todd commits to consistent delivery of exceptional cleaning, proving his reliability and value (**Specific**, **Measurable**). By excelling in his current role, Todd aims for a full-time janitorial job (**Achievable**, **Relevant**). He's set on securing it within a year (**Time-bound**). With a SMART plan, Todd increases his prospects for a position that brings with it better benefits and pay.

Here's one last example of a SMART goal: Brianna has never held a job before. Over the next three months, they're committed to submitting at least five job applications per week (**Specific**, **Measurable**), aiming to secure an entry-level position in customer service or retail (**Achievable**, **Relevant**). By setting this target, Brianna envisions themselves stepping into their first job within the next six months (**Time-bound**).

TIP

In addition to setting SMART goals for the short term, don't forget to articulate your long-term professional aspirations as context for the goals.

Put together an action plan

If using a SMART goal points you in the right direction, an action plan breaks down that goal into manageable steps to achieve the goal you set out.

Take Sophie from the previous example. She can leverage an action plan to accomplish her SMART goal by breaking down her six-month timeline into specific phases. In the first month, she could allocate time for researching suitable online courses and selecting the one aligned with her goal. During the second and third months, Sophie could focus on completing course modules, practicing design techniques, and seeking feedback. By the fourth month, she could work on creating a portfolio showcasing her skills, ready to be presented during her upcoming job interview opportunity. Regular checkpoints every two weeks could help her assess her progress and make necessary adjustments. Through this well-structured action plan, Sophie can effectively navigate her way toward her design-centered role, ensuring each step brings her closer to her desired outcome.

Todd can also utilize an action plan to achieve his SMART goal. Todd can initiate his plan by outlining a monthly schedule. In the first two months, he could focus on consistently exceeding cleaning expectations and maintaining the highest standards of cleanliness. By the third and fourth months, Todd could proactively seek opportunities to take on additional responsibilities that showcase his dedication. As the sixth month approaches, he could express his interest in a full-time role to his employer, backed by his record of exceptional performance. Regular meetings with his supervisor every two weeks could ensure that Todd's progress aligns with his goal. This plan propels Todd toward his target, allowing him to demonstrate his readiness for a full-time janitorial job.

TIP

Include in your action plans actions that help you progress toward your long-term aspirations as well!

Gather and use your resources

For action plans and SMART goals to succeed, you need the right resources. These include time, money, information, energy, and equipment. They're the tools needed to turn plans into actions. Without them, your goals stay as ideas without being realized.

Recall the example of Mia and her dream of opening a coffee shop. Mia recognizes the importance of gathering the resources needed to achieve her SMART goal. First, she plans to gather market research data and industry insights to inform her business plan. She'll also invest time in networking to build connections with local suppliers, coffee bean distributors, and potential customers. Furthermore, Mia will work on her financial acumen to secure a business loan that will fund her startup costs, including rent, equipment, and initial inventory. By marshaling these resources alongside her determination, Mia ensures that her goal can become a reality.

As for Brianna, they plan to seek guidance from career advisors and mentors who can offer valuable insights on crafting effective resumes and cover letters. Brianna also intends to tap into online job search platforms, local job fairs, and networking events to identify suitable opportunities in the customer service and retail sectors. By leveraging these resources alongside their determination, Brianna ensures that their SMART goal becomes more than a mere aspiration — it becomes a strategic plan with tangible steps toward their first job.

TIP

Understanding your values, your hopes, and your goals can help identify the types of roles that best suit you. So does understanding your personality — that pattern of thoughts, feelings, and behaviors that allow you to be you.

Monitoring your progress

Think back to the tale of how Polynesian adventurers explored the Pacific. They navigated the unknown by tracking the stars, studying ocean currents and weather systems, and observing migrating birds and marine life; they didn't just float about. There's another key element here that guaranteed the Polynesian's success: *self-examination*. In other words, they monitored their progress.

To merely set sail without tracking your course is to flirt with disaster. Monitoring your progress allows you to measure the distance from where you started and to see how far you've ventured. It also enables you to correct your course when things aren't working out. There must have been many times when Polynesian adventurers followed a current, chased a flock of birds, or sailed by how they read the wind . . . only to be wrong and led off course. That didn't mean they remained lost. They reflected on actions and made changes needed to keep pushing ahead. Like these adventurers, you can ensure that your journey is not adrift by routinely reflecting on where you are.

Here are some ways you can monitor the progress of your employment strategy:

>> **Set clear goals.** At the beginning of your journey, define what success looks like for you in the short term and the long term. By identifying these things early in the process, it becomes easier to track your progress. And a word of advice: Goals often change over time due to new information and opportunities. It's more than okay to adjust your goals as you go along.

>> **Conduct regular reviews.** Schedule periodic self-assessments to check on your progress toward your goals. You may decide that you will dedicate 30 minutes on the last Friday of each month to review your progress. Or, perhaps you want to do a deep-dive every three months to really go over your skills, go over potential opportunities to help you reach your goals, and review feedback from those who are helping you along your way. Structure regular reviews in a way that works for you.

>> **Ask for feedback.** Seek input from mentors, family members, colleagues, and friends to gain new perspectives on your growth. We all have our own perspectives, which means all of us see things that others don't. The people who want to see you thrive have valuable views to share.

>> **Track your goals.** Maintain a list of achievements, tasks completed, and skills acquired. Also keep track of any setbacks or approaches you tried that just didn't work. There's no need to fear those things. By reflecting on them, you can more easily figure out the things *that do* work for you.

>> **Track milestones.** Keep track of any promotions, raises, and major projects you've completed. This helps you highlight them when discussing potential new roles or opportunities. Celebrate and reward yourself along the way as you reach your milestones.

>> **Adjust your course if needed.** Consider whether your current direction still aligns with your goals. If you find yourself off course, adjust your goals and strategies so that they match where you want to go. Relax; adjustments are a very normal part of the journey.

By tracking and reflecting on your progress, you can showcase a strategic approach to your current managers, potential employers, and your professional network. When you take a goal-oriented approach, it highlights that you're a professional.

Nurturing your professional network

Networking is a powerhouse for professional growth. After all, it's not just about what you know, but who you know. Building connections can open doors to opportunities you may not have found. Genuine relationships and connections with others can steer your career in positive, unexpected directions.

Authenticity matters. When you network, don't put on a show. Be you. People can often feel when others are being fake, but authentic conversations have a lasting impact. Sharing your genuine stories and passions leads to meaningful connections. It's not about impressing people — it's about building relationships. So, when you're networking, be true to who you are. It's the real connections that truly count.

Being neurodivergent can actually be an asset in networking. Your authenticity and unique viewpoints can make you stand out. If you're shy or introverted, remember that being yourself is your strength. You don't always need to follow a script — your genuine approach can help you build connections with those who appreciate your perspectives and skills. Networking is about finding those who support you.

Networking can feel much more comfortable and natural when you are connecting with others who get you. For suggestions of groups of other neurodivergent professionals you may want to network with, check out the appendix.

Being authentic in networking doesn't require a grand script; it thrives on genuine actions. Here are some tips on how to be authentic as you network:

>> **Start small.** Engage in conversations that truly interest you. Ask thoughtful questions and actively listen to others.

>> **Share your experiences.** Open up about your background and experiences to foster meaningful discussions. Some neurodivergent people find it difficult to talk about themselves. However, often it can be easier to talk about your work, accomplishments, and skills.

>> **Focus on quality over quantity.** Focus on building deep connections rather than amassing a large number of contacts. With each person with whom you network, find things you have in common.

>> **Choose comfortable settings.** Opt for one-on-one meetings or online interactions if large gatherings feel overwhelming.

>> **Remember that consistency matters.** Stay in touch with your connections by sharing updates.

>> **Offer help.** Be willing to assist others when you can, nurturing reciprocity in your relationships.

>> **Expect gradual process.** Understand that networking is an ongoing journey, not a one-time event.

Exploring Career Opportunities

Exploring career opportunities often implies job searching, but it can also mean discovering new projects, responsibilities, and learning experiences within your current role. It can also mean taking advantage of leadership or skills-based courses. Exploring career opportunities is about seeking growth and challenges, both within your workplace and through broader professional avenues.

Scoping your job opportunities

As you explore career opportunities, it is important to be aware of effective job search strategies. Even if you're content in your current role, some of these tips may still benefit you:

TIP

>> **Identify different job types.** What job types align best with your strengths, values, qualifications, skills, and experiences (for example, data analyst, video producer, teacher)?

>> **Research target industries.** For the job types you identify, think about which industries are most relevant and learn about them through online and other resources such as company websites and industry magazines.

>> **Identify skills to develop.** Using a structured template, evaluate how well you match up with job descriptions and identify the gaps you need to fill (for example, certification in a specific software or an internship).

TIP

An important aspect of exploring your opportunities is to develop an understanding of how you can position yourself for the opportunities. Reviewing the qualifications of people who are currently in your desired roles and how you compare to them can be very helpful.

This comprehensive approach can help you understand the opportunities better, tailor your efforts to align more closely with your target positions, and be more prepared and confident when reaching out or applying for them.

Most qualified candidates don't perfectly align with every requirement of a job posting, and that's completely acceptable. Neurodivergent candidates are notably less likely to apply for these positions compared to their neurotypical counterparts. As Amy Root, a neurodivergent professional in Oklahoma notes in Episode 12 of the *NeuroSec Podcast,* "There's a large amount of people that opt out of certain jobs simply because we take job requirements a little bit more seriously. Like if you say you need ten years of experience in something, if we have eight, we're not going to apply to that." We encourage you to apply for those jobs you largely, if not perfectly, match.

WHERE TO SEARCH FOR A GOOD-FIT JOB

Job hunting takes persistence, effort, and a dash of grit. As they say, looking for a job is a job itself. It used to be that most job announcements were published in a newspaper. Today, information about job opportunities is presented in many forms. Here are some:

- **Company websites:** Many opportunities are first announced on a company's own website.

- **Online job boards:** Many job postings appear on such job boards as Indeed, Google Jobs, and Glassdoor.

- **LinkedIn/social media:** Many employees use these online platforms to share job openings in their company.

- **In-person applications:** In industries such as retail and construction, it's common for job solicitations to be posted onsite. In that case, *go in and apply!*

- **Recruiter or staffing agency:** Companies often use these resources to find talented individuals, especially for senior-level positions.

- **Networking:** Personal recommendations or referrals often lead to successful job placements.

- **Career fairs:** Many companies participate in recruitment events to advertise their jobs and to receive applications.

- **Competitions:** Some fields such as data analysis and coding design may hold hackathons to identify new talent that a company may want to hire.

Searching for roles

Okay, chances are you're not an Oscar-winning star exploring new career opportunities, but just like them, embracing a fresh role can lead to unexpected and exciting pages added to your life's script. But while many stars have agents to help them find new roles and navigate their careers, most likely you're doing this on your own. So, how do you effectively search for new roles? Here are some tips:

>> **Research the industries you are interested in.** Stay updated with industry trends and developments.

>> **Network.** Try to connect with professionals and recruiters in your field, as they often play a crucial role in matching candidates to jobs.

>> **Search online platforms and company websites.** Search online job sites and explore internal opportunities.

>> **Build an online presence.** Choose online platforms suitable for your profession (such as LinkedIn) and engage regularly with comments and content.

>> **Be open to feedback and remain persistent.** Accept feedback and stay patient and persistent.

Building a great resume

Alright, it is time to chat about building your resume, or CV (curriculum vitae) as it is called in some countries (such as the United Kingdom) and in some circles (such as academia). Think of your resume as a summary of your best professional moments and skills. Even if you're not job hunting, regularly updating it is smart. Why? Because unexpected opportunities can pop up anytime, and an up-to-date resume ensures that you're always ready. Moreover, revisiting and updating it can be a boost, reminding you of what you've achieved and guiding you on future goals. In the world of work, a sharp resume is a big help in your career growth.

In our courses tailored for neurodivergent job seekers, we teach our students how to read a resume as an employer. This includes looking to see whether the resume is relevant to the job the employer wants to fill, whether it is easy to understand, whether it lists the candidate's achievements, and whether it seems professional. We even invite students to critique our own resumes and we've actually updated our own based on things our students have pointed out!

TIP

As you build your own professional resume/CV, here are some tips to keep in mind:

>> **Clarity:** Use simple language and clear headings.

>> **Relevance:** Tailor content to the specific job applied for.

- >> **Quantify achievements:** Highlight your impact with numbers not just tasks (for example, "Increased sales by 15 percent"), using the STAR method (Situation, Task, Action, Result) for clarity.

- >> **Consistency:** Maintain a uniform format and font throughout.

- >> **Keywords:** Include industry-specific terms and phrases.

- >> **Action verbs:** Start bullet points with strong verbs, such as *achieved* or *managed*.

- >> **Brevity:** Limit your resume to one or two pages for most roles.

- >> **Error-free:** Proofread for spelling and grammatical mistakes.

- >> **Updated:** Regularly review and refresh your resume with recent accomplishments.

Over 98 percent of Fortune 500 companies utilize an applicant tracking system (ATS) to sift through job applications. That means that the first person to read your resume at a company may not be a person at all. Occasionally, an ATS may overlook a suitable candidate simply because the terminology on their resume doesn't align with words used in a job listing.

TIP

To improve your chances of getting your resume seen, ensure that your resume's terminology mirrors the job post. For instance, if the job listing mentions "data processing" and your resume says "data handling," consider making a change to match the listed term.

Developing a stellar profile on LinkedIn

In our digital age, a robust resume and LinkedIn profile act as your introduction to potential employers and professionals. Your resume highlights your achievements, skills, and value, guiding hiring managers in assessing your fit. The LinkedIn online professional networking platform broadens this story, offering networking opportunities and industry insights, and uncovering unexpected roles. Together, they solidify a memorable professional impression.

TIP

Following are tips to help you create a stellar online presence on LinkedIn:

- >> **Make your profile shine.** Use a professional photo, detail your work experience, and share major accomplishments achieved in each role listed.

- >> **Stay active.** Engage with relevant posts and updates by commenting on them and sharing.

>> **Network.** Connect online with professionals in your industry and those you personally know.

>> **Ask for endorsements.** Ask people you know professionally to endorse your skills (an option on the platform). Endorse the skill of others as well.

>> **Join groups.** Engage in industry-relevant groups on LinkedIn. Group members often share job opportunities, provide up-to-date industry information, and alert others of networking events.

>> **Share and publish information.** Share work-related articles you find interesting along with your thoughts.

>> **Take classes from LinkedIn Learning.** Enhance your skills with available courses available through the platform.

>> **Set alerts.** Monitor new job opportunities by setting alerts to notify you of openings.

>> **Update your profile regularly.** Refresh your profile with new achievements as you accomplish them.

REMEMBER

There is nothing wrong with getting help in establishing or updating your LinkedIn profile and presence. As an example, John on our author team shares the following story:

> When I moved to California after years of living in Washington D.C., I suddenly realized my long-neglected LinkedIn profile desperately needed to be updated (it barely had anything on it at all). Yet, I really struggled to understand how to go about it. I tried and tried, but I just couldn't get it.
>
> When I reached out to neurotypical friends for help, they all thought I was joking. You see, while working for the White House, I had twice been to LinkedIn's headquarters to meet with its senior executives. I knew more about LinkedIn *as a company* than almost anyone else. That's why my friends thought my pleading for help was a joke. Plus, I didn't yet know how to advocate for my cognitive needs.
>
> Thankfully, I had joined a networking group of autistic adults and their families in San Francico called AASCEND. A few weeks later, AASCEND partnered with LinkedIn to offer free profile-building lessons for autistic adults just like me. Thanks to a LinkedIn volunteer taking the time to show me how, I finally got it and I quickly went about building a LinkedIn profile that I think is pretty great.

TIP

You can search for jobs on LinkedIn. Makes sure to use filters for location, company, experience level, and job type to make your searches even more efficient.

Interviewing with confidence

Typically interviews are a necessary step when seeking a new career role. Whether you're applying for a new job and facing a formal interview, or aiming to talk with your manager to discuss adding responsibilities to your current position, the process often involves a discussion. It's natural to feel a bit nervous ahead of an interview. But remember, interviews are just a way for both sides to understand each other better. You wouldn't accept a job without discussing it first, right?

TIP

To interview confidently, be genuine. Everyone has unique experiences, talents, and passions. Instead of comparing yourself to others, share your distinct skills, values, and views.

REMEMBER

Most people get nervous during interviews. Some may even feel the need to oversell their skills. However, an interview isn't about bragging about the things you can do best. It's a two-way conversation to exchange ideas. Viewing it this way can help ease your nerves and give you more confidence.

Following are some tips to help you enhance your interview skills:

>> **Prepare in advance.** Research the company, its culture, and the specific role. Anticipate interview questions by analyzing job requirements.

>> **Manage your nerves.** Breathe slowly at your natural pace. Remember that the interview is just a conversation.

>> **Communicate clearly.** Speak slowly and clearly. Listen actively and don't interrupt.

>> **Focus on your strengths.** Highlight your strengths and experiences relevant to the role. Use the STAR method (Situation, Task, Action, Result) when answering questions. Show your research. Discuss accomplishments that align with the role. End with a reflection or lesson learned.

>> **Ask questions.** Prepare thoughtful questions ahead of time.

>> **Show enthusiasm.** Show your genuine interest in the company and the role. No need to fake enthusiasm.

>> **Be authentic.** Be honest about your skills and experience. Don't exaggerate.

>> **Follow up.** Send a thank-you email afterward to those you interviewed with. Keep it brief.

Being Effective at Work

Once you have a job, you can't rest. Work is *work*, after all. As with any job, if you are unfocused and unprepared, the job of *work* can become very difficult. Here, we look at approaches that can help you thrive and be your best in your work.

Exercising a solid work ethic

Having a good work ethic means more than just doing hard work; it also defined by how you approach and do tasks, and it includes your attitude and behavior toward how you work. Key elements include:

>> **Reliability:** Being punctual and meeting commitments.

>> **Dedication:** Committing to tasks, especially during challenges. Maintaining a positive mindset, even when you face challenges or setbacks. Showing initiative and not waiting to be told what to do.

>> **Productivity:** Having clear objectives, keeping track of your tasks and responsibilities, and efficiently completing tasks with quality.

>> **Cooperation:** Collaborating well and fostering a positive work environment. Staying on top of your email and calendar. Regularly asking for feedback from your boss and colleagues.

>> **Integrity:** Acting honestly and ethically, even when no one is watching.

REMEMBER

Building a strong work ethic involves setting personal standards and practicing consistency in those standards over time. While a strong work ethic is an overall pattern of behavior, *we all have our bad days!* Keep at it.

Being assertive in your conversations

Communication is a vital part of our work lives. No matter the job, we'll have coworkers with whom we'll need to communicate. Think about it: Even a solitary lighthouse keeper on a remote island has coworkers. Sure, they may not share lunch breaks, but those voices that crackle back on the radio? Or, the folks in the main office reading the written-out lighthouse reports? Those folks are coworkers too.

Understanding common workplace communication styles

Think about the conversations you have at work, especially those times when there is stress. Does it often feel as though you just can't get your point across? Do conversations with certain people always feel like a battlefield? Do you over worry about offending others, or have a hard time speaking up for yourself? Each of those things are expressions of common communication styles in the workplace; almost none of them helpful. Following are a few of those communication styles you may recognize:

>> **Passive:** Holding back opinions or letting others override your rights? This can result in resentment or anxiety because your needs go unmet.

>> **Aggressive:** Expressing feelings strongly, even at the expense of others? This can breed fear or hostility, distancing you from people.

>> **Passive-aggressive:** Seemingly easy-going but with underlying resentment? This hinders your personal growth.

>> **Assertive:** *Our aim!* Clearly express feelings and needs without overstepping others' boundaries. This fosters mutual respect and growth.

When people hear "assertive communication" they sometimes mistakenly confuse it with aggressive communication. Here's an example that shows how the two styles are really quite different:

> Pedro and Elena are coworkers in the same department. When faced with a project deadline, Pedro, with his aggressive communication style, often says things such as, "You better have this done by tomorrow, or it's on you." He believes that by exerting pressure, he can get things done faster. Elena, on the other hand, opts for an assertive approach. She communicates by saying, "I understand we're under a tight schedule. How can we work together to meet the deadline?" While Pedro's approach may cause tension and stress, Elena's fosters collaboration and understanding. Over time, it's evident that Elena's assertive style not only maintains a harmonious working relationship but also leads to more efficient and productive outcomes.

Assertive communication involves expressing thoughts, feelings, and needs in an open, honest, and respectful way. This style promotes understanding, fosters positive relationships, and ensures that everyone's rights are acknowledged and valued.

Take a moment to reflect on your own dominant communication style. Is it passive, aggressive, passive-aggressive, or assertive? Recognize that various factors contribute to your dominant style, such as past trauma, personality traits, and perceptions.

A person's communication style isn't fixed. Assertive communication can be developed with practice.

Developing an assertive communication style

So, how do you develop an assertive communication style? Well, through intent (for yourself and the other person) and practice (and more practice). It's like doing yoga, or going to the gym, or taking up surfing. You don't master it right away, but keep at it and you'll eventually get there. Following are a few practical tips to keep in mind:

>> **Awareness:** Recognize and understand your own feelings and needs. Aim for a mutual understanding and alignment with the other person. Be open to modifying your intentions and approach.

>> **Active listening:** Pay full attention to the other person without interrupting. It's okay to check to make sure you've understood things correctly.

>> **Clear language:** State what you're hoping to achieve, describe any relevant data, give your reasons, and round off with your conclusions. Be specific and straightforward in your messages. Frame thoughts as "I feel" or "I need," rather than blaming or accusing. Clearly define your limits and communicate them respectfully.

>> **Practice calmness:** Stay calm, even when the topic is challenging or emotional.

>> **Seek feedback:** Ask trusted individuals about how they perceive your communication style.

When faced with diverse communication styles in the workplace, assertiveness is key. Encourage passive communicators to express themselves. With aggressive individuals, remain calm and respectful. If they remain unconstructive, consider involving managers.

Resolving conflicts with confidence

Whenever there is a disagreement between opposing ideas, interests, or individuals, we call that *conflict*. And while many people may think of war when they think of conflict, conflict is actually an everyday part of life. In fact, it's a fundamental aspect of being human, given that we each have our distinct perspectives and backgrounds.

In workplaces, differences can sometimes cause friction. While this can spark new ideas, managing conflict is essential for collaboration and harmonious coexistence. Here's an example of conflict inside the workplace:

> At the marketing department of a tech company, two team members, Kiara and Eric, disagreed vehemently on the direction of an upcoming campaign. Kiara believed a futuristic, tech-forward theme would resonate more with their younger audience, while Eric was convinced that a nostalgic approach would differentiate them in a saturated market. The escalating tension began to disrupt team meetings and hinder progress. To address the situation, their manager arranged a brainstorming session where both Kiara and Eric can present data supporting their viewpoints. By focusing on concrete information rather than subjective opinions, the team collaboratively devised a strategy that combined elements from both ideas, ensuring a unique campaign that appealed to their target demographic.

Looking at common approaches to conflict in the workplace

Resolving conflicts effectively involves several strategies, not all of them effective.

>> **Avoidance** means sidestepping disputes. Useful when the timing isn't right but risky as unaddressed issues can escalate.

>> **Accommodation** values relationships over disagreements, emphasizing harmony, yet may leave you feeling shortchanged.

>> **Compromise** seeks middle ground, balancing quick solutions with potential dissatisfaction, but often no one ends up happy.

>> The **competing** approach views conflicts as battles to win, which may be effective in urgent situations but may strain relationships.

>> **Cooperation** frames conflicts as opportunities for collective solutions, prioritizing long-term trust and collaboration but demands patience and engagement from all parties.

Which approach feels most like you?

Understanding how to resolve conflict

Quickly resolving conflicts at work helps everyone, but few are ever taught *how*. Remember Kiara and Eric from the earlier example? The resolution of their conflict demonstrated several of the following foundational principles that you can use to solve workplace conflict as well:

- >> **Acknowledge and address:** Own up to your part. Were your words or actions perhaps a catalyst? Recognize the needs and perspectives of everyone.

- >> **Destination in mind:** Keep your bigger goals front and center. Choose the approach that aligns with them. Avoid escalating tension, including using language that accuses others.

- >> **Keep it essential:** Identify the outcomes you want. Is it urgent? Important? Both?

- >> **Mutual wins:** Aim for balanced resolutions. Be open to new solutions, even if it means letting go of your initial stance.

- >> **Listen, then speak:** Always ensure that you truly understand the other person's viewpoint before diving into yours. Promote honest dialogue.

- >> **Collaborate:** Clearly define the issue at the center of the conflict. In most conflicts, people usually agree on some things. Start there. Engage in brainstorming sessions and savor both the process and the results. Figure out a solution that works for all — often it's an idea no one has yet considered.

- >> **Self-care:** Don't forget to relax and recharge. Conflict resolution can be taxing.

Working well with your team

Organizations revolve around missions and goals. Picture it: The organization is an orchestra, with each team as an instrument section. Every team must align with the orchestra's main tune. To excel in your job and advance, it's crucial to align with your team's goals and understand the organization's mission and structure. Here's how:

- >> **Be clear on your role.** Knowing your position in the organization is essential. Ask yourself: Which team are you a part of? Who guides your team? What are your role and responsibilities? Understanding these can better position you within your team and the broader organizational structure.

- >> **Maximize meeting effectiveness.** Want to ace meetings? Show up on time, be prepared, and most important, be present. If you're facilitating, ensure that everyone's voice is heard, stick to the agenda, and wrap up punctually.

- >> **Work the team dynamics.** Collaboration and communication are the heartbeats of any team. Workplaces may use online platforms (such as Google Workspace or Slack) to foster collaboration, while others may stick to traditional meetings and talking face to face. Whatever the method, the key to successful collaboration is to be clear in your communication.

Giving and receiving feedback

Effective feedback drives personal and professional growth. While challenging, it's essential in work. Here's how to receive and give feedback effectively:

>> **Receiving feedback:** Actively listen without interruptions. Stay open and seek clarification. Reflect on and act upon feedback. Revisit feedback and apply or discuss further. Accept feedback as something that helps you grow.

>> **Giving feedback:** Provide feedback promptly and offer ongoing support. Be specific and avoid excessive critique. Address actions, not the individual. Use a balanced approach and avoid highly critical terms. Remember, it's just your perspective.

Evaluating your performance

Performing well in a job starts with setting goals with your manager and applying all your workplace skills to perform to the best of your abilities to meet those goals. Effective communication with your manager is crucial for enhancing your performance. Regular check-ins ensure alignment with expectations and identify improvement areas. Actively seeking your manager's feedback shows dedication to growth and nurtures a relationship of mutual respect and continuous learning. This partnership benefits both current tasks and future career direction.

REMEMBER

Feedback and performance reviews play a pivotal role in your growth journey. Consider the feedback you receive from your colleagues and managers as a gift.

Continuing to learn

Embracing continuous learning is vital for your professional growth. In a swiftly changing business world, keeping up with the latest trends and skills ensures that you stay relevant and ahead of the curve. By adopting a lifelong learning mindset, you can enhance your expertise, foster creativity, and be more adaptive. This dedication to growth can open doors to new opportunities and career advancements, ensuring you enjoy a richer, more fulfilling professional journey.

TIP

Many companies many pay for or help subsidize training programs that can teach you new skills and develop your leadership abilities. Some companies may also pay dues for professional organizations or conference costs for events that focus on professional development. This isn't always widely advertised, *so ask!*

Becoming a Great Employee

Understanding how to navigate a neurotypical workplace as a neurodivergent individual is important. It's not about conforming but valuing who you are.

Embracing your normalcy

Being neurodivergent at work is just like bringing any other aspect of yourself to work. There's no need to shy away from having normal variations in how you think and experience the world. You can't change those things about you. What you can do is connect with the support you need to find success.

Seeking accommodations

Earlier in this book, we talk about how every single human uses accommodations in their everyday life. They aren't just limited to neurodivergent or disabled people. (Don't believe us? Check out Chapter 2 for more.) There can be a lot of fear in asking your employer for accommodations. But, in the end you're simply asking for the things you need to do your job well.

Connecting with employee resource groups

In 2018, a group of employees at Block, a company known for its Square and CashApp products, initiated an employee resource group (ERG) for neurodivergent employees. Starting with about ten members, it now has over 2,000. There is a benefit to connecting with other employees who get you. If your company doesn't have a neurodiversity ERG, look to disability ERGs and other employee groups for support. In addition, you can tap into global networks of neurodivergent professionals.

Being authentic

No one teaches you how to be an authentic, neurodivergent employee — that's one reason why we wrote this book! We often need to trade tips and tricks with others to help us work our best. Here are a few helpful ones:

>> **Self-awareness:** Often, we're aware of our own needs as neurodivergent people. It sounds simple, but paying attention to what you need makes a big difference.

>> **Self-advocacy:** Representing and asserting your needs, especially regarding your neurodivergence, is normal. There's no shame in advocating for them!

- >> **Empathy for managers:** Like you, managers have their own priorities and needs. Regularly check in with your managers to align your tasks with their goals. This sincere approach can greatly enhance your professional relationship.

- >> **Examples:** Many neurotypes think expansively, aiding in strategic planning but sometimes making it difficult to start tasks. A solution is seeking examples. If your manager assigns you to write a memo you're unfamiliar with doing, ask for a similar one they've found spectacular. An example not only focuses your thoughts, but clarifies your manager's needs.

- >> **Timers:** Timers can enhance focus. For large projects, set a timer (for example, 6, 12, 16 minutes) and see how far you progress within that span. It's often surprising!

- >> **Deadlines:** Many neurodivergent people excel when nearing a deadline. Instead of waiting for major deadlines, break tasks into smaller segments with quicker due dates.

- >> **Downtime:** Schedule regular breaks to manage sensory overload or reduce stress. Even short pauses can recharge your focus.

REMEMBER

You don't have to tackle everything by yourself. It's totally okay to reach out for support from others. If you need help, just ask for it!

IN THIS CHAPTER

» **Strengthening personal relationships**

» **Navigating the dating world**

» **Parenting neurotypical and neurodivergent children**

» **Finding community and connecting with friends**

» **Accessing services, accommodations, and health care**

Chapter **16**

Thriving in Relationships and Communities

Relationships are the threads that connect us all, and they're just as vital for neurodivergent individuals. Whether it's family, partners, friends, or community, we all need support and connection. But let's be real. Understanding how to heal and grow in relationships is no small feat. It's not like they teach this stuff in school. While some neurotypical people may naturally grasp conflict resolution and connection, it may not be so straightforward for neurodivergent individuals.

So, what's the game plan? In this chapter, we dive into how neurodivergent individuals can develop and understand interpersonal relationships. We explore the importance of community, support, navigating health care, and standing up for one's rights. It's about connecting the dots, finding balance, and recognizing that we're all in this together, learning and growing.

Thriving in Personal Relationships

Personal relationships — whether friends or family — are a foundational part of our lives. These relationships feed our souls, giving us emotional backing, a place to belong, and a boost to our feeling of self-worth. They influence everything — our decision-making, how we bounce back from tough times, the way we connect with others, and our happiness. Our relationships shape our world and how we find meaning in it.

But it's not just about *feeling* connected. Romantic partners, family members, and friends all have a unique impact on us, and it's crucial to recognize that we are shaped by the people around us — and in turn we shape them. Finding ways to strengthen these relationships can help us find that sweet spot of balance and harmony.

For those of us who are neurodivergent, strengthening our relationships is even more essential. We live in a world of people who don't fully get us, and that can be challenging. Like every other human, we need relationships that we can retreat to, feel safe in, help us grow, and make us feel understood. At times, we may need to help our loved ones understand us. After all, they can't read our minds. But we also need to put in the effort to understand them. It's not just about connection; it's about creating a life that's fuller and more satisfying for all of us.

Even if relationships in your life seem broken, it is possible for them to heal. Understanding the other person's perspective is the first step.

REMEMBER

If there is any form of abuse — physical, emotional, or sexual — in a relationship, it is a perfectly valid choice for you to disconnect from that relationship.

WARNING

Our parents and caregivers

Relationships with your parents and caregivers are complex and have lasting impacts on your life, affecting how you relate to others in the future. It's essential to make an effort to understand and nurture these connections whenever possible.

Understanding our parents and caregivers

Have you ever felt that your parents just don't get you? (If you answered no, then you must have never been a teenager.) It's normal for humans to feel disconnected from their parents at various stages of life. However, for many of us neurodivergent folks, this feeling of disconnect is something that we may particularly wrestle with — especially if we're the only neurodivergent person in our family.

Perhaps you feel that your parents aren't tuned into your struggles, or that they can't really understand all the things you are going through. Maybe you feel invisible. Perhaps you feel unheard despite all your efforts. Maybe you've given up, or maybe you try and try and yet nothing seems to work.

But here's a thought: Have you truly tried to understand your parents? You may have heard the saying, "parenting never comes with a handbook," and as pat and annoying as that may be to hear, this is truly the case. None of our parents automatically understand neurodivergent conditions or how to raise a neurodivergent child. On top of that, we can almost guarantee you that no doctor, therapist, or educator gave your parents the information they needed to fully understand you.

Society has done a horrible job of supporting the parents of neurodivergent children (more on that in Chapter 19). So, they don't automatically know how we experience the world and how our experiences may differ from their own. We've got to help them out, and that begins with a little empathy and patience.

REMEMBER

If you want your parents to understand you, it's important to understand all you can about yourself first. Self-reflection and acceptance of who you are is key to that.

TIP

Communicating how you experience a situation, or how something makes you feel, is like unlocking a door to better understanding. Need to talk without interruptions? Just ask. Want a peaceful chat where you can also grasp your parent's or caregiver's viewpoint? Lay down that groundwork. But here's the golden nugget: To truly get someone else, you've got to listen, *really listen*, to what they're going through and try your best to walk in their shoes. Hear what others have to say and try to grasp their perspective. If you do, you'll find that empathy opens doors. It's not flashy, but it's how real understanding happens.

Listening can be a powerful tool. By truly tuning in, you may discover something you hadn't thought of before, or realize that your words and actions landed differently than you meant them to. In conversations like these, every insight can be a step toward a more robust relationship with your parents. But listening is not just about nodding your head; it's about genuinely absorbing what they have to say, taking it to heart, and using it to shape how you interact with them moving forward. Developing a more robust relationship may also be about speaking more clearly, being more conscious of how you come across, or making a clear commitment to enhance the relationship. It's about giving everyone a chance to feel seen, heard, and understood. And that's just common sense mingled with a bit of kindness.

One of the biggest reasons we may find it hard to understand our parents is because they may not agree with our choices or our decisions. Our parents may state that they want what is best for us. Sometimes what's best in their opinion may differ from ours, and over time we may feel angry or resentful of their feedback and be unwilling to hear anything they have to say. Guess what? *Every* human goes through that!

If we can grasp the concept that our parents just want us to be happy, we can demonstrate to them that we know ourselves best and that our choices and decisions are truly those that make us happy. If we can get that they're concerned about our future, we can communicate our dedication to our goals. We can tell them that, yes, we're worried about our future too, but we're committed to growing, to improving our skills, and to striving for a better version of ourselves. It's about opening up a dialogue and finding that common ground.

Look, being a parent is almost an impossible task, and especially for parents of neurodivergent individuals. Figuring out how to support and protect us can feel like trying to solve an unsolvable riddle. Many parents operate from a place of fear, wanting to keep us safe. But here's where we can help: We've got to stand up for ourselves, showing them that we can be independent and self-sufficient. Sometimes, we may feel as though our choices are being overshadowed by what our parents think is right. But opening up, communicating, and demonstrating our capabilities can help forge a path that embraces both our autonomy and their understanding. It's a balancing act, but it's one we can achieve together.

REMEMBER

Working with your parents instead of against them can lead to better communication of your needs and perspectives, ultimately enhancing your relationship.

Nurturing our parents and caregivers

An essential aspect of relationships is the duty to nurture each other, which means providing the care, encouragement, and support that helps a loved one grow and thrive. Yet, if you search online for information on how someone can nurture their parents, almost every item that comes up talks about how parents can nurture their children. Perhaps that's just the nature of parenting. Our society places so much emphasis on parents nurturing their children that many of us don't even consider that our parents need nurturing too.

Khushboo from our author team often shares how many parents of neurodivergent children face isolation from their communities due to misunderstandings or stigma. They may feel disconnected, saddened, or unaccepted by others who don't understand or embrace their child's neurodivergence. These parents often soldier on, focusing on being strong for their child, even as they may feel lonely or burnt out. Even if they are not your own parents, if you know someone in this situation, take time to nurture them when you can. Simply asking how you can help or offering small gestures of support, such as assisting with chores or introducing a new hobby, can make a significant difference. Sometimes, just being present and offering a kind word can nurture the relationship and alleviate some stress.

Nurturing our relationships with our parents and caregivers requires an active effort to practice and maintain respect, compromise, commitment, and mutual benefit.

Understand what respect, compromise, commitment, and mutual benefit mean for both you and your parents. Ask questions, make efforts to improve, and find areas to compromise. Communicate your boundaries and follow through on promises. This approach can lead to healing and progress in your parental relationships.

Our spouses and partners

Relationships can be complex, and this holds true for neurodivergent individuals and their partners or spouses. Communication, often a challenge in any relationship, can be even more so between a neurodivergent person and their partner because of differences in communication styles. Even among two neurodivergent partners, those styles can vary widely. But what's universal in any relationship is the need to truly understand our partners. We must take the time to listen, grasp what they need from us, and consistently put in the effort to nurture that connection. Nurturing this relationship is about learning, growing, and fostering the bond we share, one careful step at a time.

Understanding our spouses and partners

Self-awareness, particularly for neurodivergent individuals, is a doorway to understanding our partners better. The more we recognize and embrace our own quirks and unique characteristics, the more we can appreciate those of our partner. It's not about expecting perfection, but rather recognizing that we don't intentionally hurt one another, and that compromise and understanding are two-way streets.

In any relationship, there will be differences. Maybe those differences are small, like a preference for wearing slippers around the house versus going barefoot, or perhaps they are more significant, like how each person handles conflict. Food preferences may clash, or there may be misunderstandings about feelings and behaviors. The reality is, finding a person exactly like you is an impossible task (and where is the fun in having a partner exactly like you?), but that doesn't mean harmonious partnership is unattainable.

What really matters is finding common ground amidst the differences. It's about working together to identify those shared values, goals, and rules that can make the relationship thrive. This involves recognizing what you can compromise on and what is nonnegotiable, all while cultivating a healthy, safe, and nurturing relationship. And in the end, isn't that what we're all striving for?

The best way to understand your partner is to listen to them.

Listening to your partner is more than just a courtesy; it's a way to show you genuinely care. It's about asking the right questions and understanding their love language, their needs, what they like and dislike, and what makes them feel cared for. It's about delving into what really matters in the relationship, even when that means facing the things that are frustrating or annoying.

Yes, this can be hard. But just the act of listening can work wonders in healing a connection and showing a real commitment to work on the relationship. Listening is not just about hearing, but also reflecting back what you've heard and validating your partner's feelings. It's about keeping that defensive wall down and not shutting off, waiting patiently for them to finish before sharing your perspective.

And when it's your turn to speak, be clear on why you may see things differently and factor in their view to find a solution that works for both of you. It's a conversation, a genuine engagement that builds trust, understanding, and a stronger relationship. It's not always easy, but it's always worth it.

Nurturing our spouses and partners

Nurturing our spouses and partners calls for open communication and treating each other with kindness, respect, and consideration. It's about spending quality time together, bonding through shared activities, and embracing life's changes together. Communication, even over-communication, keeps you both on the same page as you navigate life's ups and downs. Frequent check-ins to discuss what's working and what needs adjustment can prevent minor issues from becoming major problems.

The key to sustaining a relationship is ongoing curiosity about your partner, learning how to support them in the ways they need, respecting their independence, and remembering that conflict can lead to deeper understanding if approached with consideration. It's a continuous effort to balance and grow the connection, always with an eye toward mutual benefit.

Think of you and your partner as teammates. Working together with this mindset can help you overcome any challenges that come in the way of your relationship.

Nurturing relationships with spouses or partners is a unique endeavor for each couple. It's all about discovering what makes the other person feel cared for and supported. Perhaps your partner needs time alone, assistance with chores, or a little extra affection and attention. It may not be immediately obvious, so having an open and honest conversation about what nurturing means to them can open the door to understanding how to best provide that support. By actively engaging with each other's needs, you can create a strong foundation for a nurturing relationship.

Dating as a Neurodivergent Person

Dating as a neurodivergent individual comes with its unique challenges and nuances. But hey, we're living in a time when dating is unlike what it used to be for anyone!

Self-awareness is your best friend here. You don't have to pretend to be someone else, nor should you. Be open with your dates about your preferences and any neurodivergent characteristics that may influence the dating experience. Maybe crowded places aren't your scene, or perhaps you lean more toward the introverted side. Share with them what daily routines are important to you, how you prefer to communicate, any sensory sensitivities, your interests and passions, and any personal boundaries and needs. Whatever it is, laying it out from the start helps you assess compatibility with your date more quickly. Don't forget to ask them about their own characteristics, needs, and preferences as well.

REMEMBER

This approach does more than just filter out incompatible matches; it empowers you. By understanding yourself and what you want, you can more effectively advocate for your relationship needs and desires. So go ahead, embrace your uniqueness, and let that guide you to connections that truly resonate with who you are. It's not just about finding someone who fits with you, but finding someone who celebrates you.

In the dating world, expect ups and downs, so stay grounded and don't be too hard on yourself. Embrace rejection as a normal part of the process, knowing that not everyone will be the right fit. If a connection doesn't work out, take time to heal and reassess. By remaining open-minded and willing to adjust, you'll learn more about yourself and what you want from a relationship. Whether you discover patterns that need changing or realize that you may not be ready to date at the current time, the key to enjoying dating process is to be honest with yourself. Trust your instincts and seek support from friends. It's all part of the journey to find what fits best for you.

Our Children: Parenting as a Neurodivergent Person

Being a parent is already a whirlwind of responsibilities, joys, challenges, and surprises. But what if you're a neurodivergent parent? Maybe you've known about your neurodivergence for a while, or perhaps it's something you've only recently started to understand, possibly even after your child was diagnosed with a neurodivergent condition. Either way, layering neurodivergence on top of parenting can indeed add complexity to the role.

It's not just about managing daily tasks and responsibilities. The complexity also comes from having to navigating a world that may not always be structured in a way that works best for you or your child. But don't worry; understanding your neurodivergence can also be a strength. It's an opportunity to connect, empathize, and create a parenting strategy that's as unique and dynamic as you are. In the following sections, we dig into what this means and how you can leverage your neurodivergent qualities in the rewarding journey of parenting.

Parenting our neurotypical children

Being a neurodivergent parent to a neurotypical child may mean you feel as if you're working double-time to be there in the ways your child needs. Questions such as "Am I doing enough?" or worries about how you could do more may weigh on you. If this struggle starts to feel overwhelming, it may be a good time to pause and reassess.

Think about what can improve the situation. Maybe you need help with executive functioning, or perhaps more shared leisure activities would strengthen your bond with your child. Recognize patterns in misunderstandings and consider ways to mitigate them, such as visual reminders or focusing on being more present during conversations. And don't hesitate to discuss with your child how you prefer to communicate and advocate for your own needs.

Nurturing your relationship with your child means knowing your triggers and what may lead to sensory overload. Keep the tools and strategies you need handy to feel better in those moments. Stay open to the perspectives and feelings of others and be willing to adapt for better outcomes. Embrace the unique individuality of your child, reducing any divide through understanding, acceptance, and open communication.

Provide affection in your own unique way, offer positive reinforcement, and frequently check in with your child. Mistakes will happen, and there may be times when you feel you're getting it wrong, but don't worry. Keep working on the challenges and applying feedback from loved ones to grow and develop. With the right mindset and intentions, you'll find that everything falls into place over time.

Parenting our neurodivergent children

As a neurodivergent parent with a neurodivergent child, you share an inherent understanding of one another since you both share neurodivergence. However, differences between you still exist, whether it's in communication, energy levels, or sensory preferences. Figuring out boundaries and areas of compromise is essential so that everyone feels respected and their needs considered.

Challenges may arise, and solutions may seem elusive, but with time, creativity, and collaboration, you're sure to find a mutually beneficial way forward. It's a team effort, and the key to a happy and healthy family environment is everyone working together. Remember, your shared experience provides a unique insight that can foster connection and growth in your relationship with your child.

REMEMBER

Recognizing that everyone, whether neurodivergent or not, has unique strengths and ways of processing things helps us see that differences can exist between any parent and child. This is a totally normal thing.

As a neurodivergent parent, your openness and understanding of yourself can be a guiding light for your child. Embrace your lived experience and teach your child the power of self-advocacy and self-love. Encourage them to be true to themselves and use your own journey to inspire them. Even though the world may not be designed for neurodivergent individuals, your love, wisdom, and patience can help your child thrive.

Listen to your child, even when their experiences differ from yours. Help them grow up feeling empowered and connected. You have the chance to impart wisdom that helps them see their own beauty and potential. Love them wholeheartedly and be patient and compassionate with yourself as well.

Remember to take care of your own needs, whether that means seeking professional help or leaning on trusted friends and family. Life is a learning process, and as you develop strategies for nurturing both yourself and your child, you'll create a relationship built on mutual respect and understanding. You're not expected to be perfect; you're expected to be human, and that's a powerful gift to give your child.

Thriving in Friendships

For many neurodivergent individuals, friendships can become a chosen family. These connections often provide more than just companionship; they offer emotional support, unbiased advice, and a sense of belonging. Friends can be a safe haven, understanding and accepting us in ways that others may not. These relationships often prove vital in navigating life's challenges, fostering a shared sense of identity, and promoting a genuine sense of community. Building and maintaining these friendships can enrich our lives in meaningful ways and provide the connection we all need as human beings.

Seeking friendships

For many neurodivergent individuals, building friendships may look different compared to neurotypical people. While societal norms often depict forming friendships as easier in childhood, Khushboo observes that the opposite may be true for many neurodivergent people.

During younger years, neurodivergent children may face exclusion or harassment due to their differences, making the formation of friendships more difficult. Unaware of their uniqueness or feeling isolated, they may struggle to connect with peers. However, the situation often changes positively in later years.

As they grow and embrace hobbies, intense interests, and activities that align with their passions, neurodivergent individuals can find empowerment and connection. Participating in clubs, sports, or extracurricular activities related to their interests fosters growth and self-acceptance. In adult years, connections can blossom through community involvement, hobbies, work, and self-advocacy. By educating others about their differences and leveraging their strengths, neurodivergent adults can form fulfilling friendships with those who appreciate and accept them for who they are.

TIP

For many neurodivergent people, bonding over shared interests and experiences is easier than bonding over small talk. If you meet someone with a connection to something you enjoy, explore that! Some of the strongest neurodivergent-neurotypical friendships are made this way.

The experience of feeling excluded, disrespected, or judged is a painful one that is shared by many, not only by neurodivergent individuals. Building authentic connections requires recognizing the importance of friends who make you feel safe and understood. For neurodivergent people, awareness of personal friendship goals and expectations in relationships can guide them in forming meaningful connections. It's vital to be honest about your unique characteristics and be true to yourself rather than trying to fit into someone else's mold. If you find that your expectations in a friendship aren't met, it's absolutely acceptable to seek connections that align better with who you are. The friends who are right for you bring understanding, support, patience, and compassion. Building good, lasting friendships may take time and patience, but the effort is worthwhile when it leads to connections that truly resonate with your authentic self.

REMEMBER

It's important to recognize that no one friendship is like any other. People are unique and different in how they relate to others and what they expect from friendships.

Embrace patience and a genuine approach to forming connections, and you'll discover individuals with whom you share true friendship chemistry. Once you find these connections, the journey of becoming a supportive friend to them begins.

Understanding and accepting those friends different than us

Friendships are a fascinating mix of differences and commonalities. People you connect with may see the world in unique ways, process information differently, or communicate in their own styles. And that's totally okay! The cool thing about friendships is that they can open your mind to fresh perspectives and diverse viewpoints. While not every difference is easily resolved, having shared friendship goals — such as staying connected, offering support, or having fun chats — can help bridge the gaps.

When it comes to being friends, being upfront about your needs is key. If certain things trigger challenges for you, it's smart to communicate them in advance. Honesty about your neurodivergent needs is a powerful thing. If you're having a rough time and need space to recharge, you can let your friends know and ask for their understanding. Consider the following:

> Deepa and Davina, who were high school friends, went to different colleges with different levels of competition and pressure. Deepa was upset and a bit angry because Davina didn't reply to many of her texts, making her feel ignored. However, when they met during the holidays, Davina explained how busy she was with her stressful school work. After a sincere talk, they agreed on how often they would call each other to maintain their long-distance friendship.

REMEMBER

Remember, it's fine to ask your friends to try new things or suggest doing different activities together. If you prefer hanging out with just one friend at a time instead of in a group, feel free to mention that. Here's an example of how that can work:

> Jason and Avery work together. They recently finished a project where their skills complimented each other, and they enjoyed working as a team. Avery asked Jason to hang out with him and some friends at a bar. But Jason felt uncomfortable with the loud and busy environment there, so he left early. The next week, Jason told Avery about his sensitivity to such environments. Avery appreciated knowing this and promised that next time, they would choose a place that's better for Jason.

True friends who care about you will get where you're coming from, respect your boundaries, and support you in ways that feel comfy. So go ahead and navigate those friendship waters, respecting yourself and others along the way!

Nurturing friendships

Nurturing friendships involves clear communication and grasping each other's needs. Every connection is unique, and it's important to know what works best for your friends. Take online memes, for instance — some pals may be all about constant meme exchanges, while you may lean toward quality one-on-one time every so often. It's all about finding that sweet spot where both you and your friend are comfortable and getting what you need from the friendship. So, keep those lines of communication open and tailor your interactions to fit what works best for each of you!

TIP

You may not always be able to nurture all your friendships in the ways that others may expect, but if you're open and honest about who you are and how you are able to show up as a friend, you may discover that people will respect your honesty and vulnerability and work with you to make this connection work.

Treat your friends with kindness, compassion, and a good dose of understanding. If you're seeking advice or assistance, don't hesitate to open up to friends you trust. Engage in their interests, find common ground, and cheer them on in chasing their dreams. Remember, respect is key. Instead of trying to change your pals to match your mold, embrace their uniqueness and meet them where they are.

Appreciating those differences and distinctive qualities is crucial. Challenge yourself to view things from new angles and with fresh eyes. Trust and respect are the foundation, and there's value in keeping up with communication, shared activities, and personal meetups. Emotional support and mutual growth are essential too. Keep in mind that each friendship has its own dynamics, so take time to chat with your pals and understand how you can best nurture your unique connections. It's all about building a supportive and enriching network of awesome humans!

Thriving in Communities

Being part of a community can be a tremendous wellspring of emotional support if you are neurodivergent. Communities can help foster a strong sense of connection and offer resilience-building, assistance with challenges, and a chance to learn from others who resonate with your values, culture, and interests. And these communities come in various flavors — educational, professional, online, local, cultural, artistic, activist-driven, and even identity-rooted. Regardless of the context, these communities can play an important role in making neurodivergent individuals feel acknowledged and valued. They pave the way for establishing and nurturing richer, more profound connections with others.

Why you need community

A community is a group of people who share common characteristics, such as attitudes, interests, or goals, and live together within a broader society. While not everyone may feel the need to be part of a community, as a neurodivergent individual, it can serve as a space where you can discover greater understanding and acceptance, especially when the community's values resonate with your own.

Being part of a community is like finding your people in a world that sometimes feels a bit wonky. For neurodivergent folks, these groups can be like a safe haven where you're understood without even having to explain yourself. It's like having a bunch of pals who get you on a deep level because they're all vibing with similar attitudes, interests, or goals. So, if you're neurodivergent, remember that these communities aren't just groups — they're places where you can be yourself, learn from others, and feel that awesome sense of belonging.

REMEMBER

Being a part of communities can open up doors to knowledge, tools, connections, and maybe even cool chances you may not have stumbled upon otherwise. They're like a treasure trove of information and support.

The benefit of neurodivergent community

Finding the people you click with is essential. So, it makes sense that engaging with other neurodivergent people is pretty beneficial. There are so many things about yourself that you don't have to explain when you're around others who are similar to you. John from our author team shares the following anecdote:

> I joke that sensory-friendly rooms at autism conferences should be renamed "Autistic Happy Hour." It's amazing how our communication and socialization changes when autistic people hang out together. It's not that we're less "autistic"; we're just more relaxed and don't have to waste energy translating between ourselves and a neurodivergent world.

Lots of people benefit from being around others just like them. Neurodivergent people aren't the only ones who have to code switch between two worlds. The term *code switching* was popularized among Black Americans as a way to describe how communities socialized with each other versus the wider world. If you've ever been an "other" in a majority of people not like you, you've probably done this too.

TIP

All sorts of neurodivergent communities are around that focus on various neurodivergent conditions as well as lived experiences. There are groups for professionals, people of color, women, LGBTQ+ people, those of religious faith, students, sports fans, and so many more. (Check out the appendix to find a list of neurodivergent communities to connect with.)

Accessing services

It is common for people to shy away from asking for help or support. But you know what? Getting help and support when you need it is a sign of maturity. In the following sections, we talk about some services that can benefit neurodivergent people. (You can find more information on where to connect with them in the appendix.)

Vocational

For a lot of neurodivergent folks, finding and keeping a job can be a real challenge. Sadly, many workplaces aren't clued in about neurodiversity or set up to welcome all those unique traits. That's where vocational support steps in. Neurodivergent individuals can tap into resources such as regional centers, vocational rehabilitation offices (sometimes called the department of rehabilitation), disability employment programs, disability hiring events, social services groups, and nonprofits that focus on job support. You can also explore specialized courses or coaching to pick up the skills needed to land and hold onto a fulfilling job that's just right for you.

Rehabilitation

Many neurodivergent individuals benefit from rehabilitation services that can help them with physical or cognitive challenges or simply catch up on skills that schools failed to adequately teach them. Such services include speech therapy, occupational therapy, executive functioning coaching, organizational help, and other areas where someone has a need.

WARNING

Seek out neurodiversity-centered approaches. When service providers use words such as *fix, cure, challenging behaviors, compliance,* and *normalization,* that's a pretty big red flag that indicates that their services may not be for you.

Nowadays, it is completely normal to hire services to do your shopping, walk your dog, clean your home, get organized, drive you to meet friends, or package-up your weekday dinners. What's so different about hiring a service provider to help you with the additional things you need? Bonus: In many places, your local government may pick up the tab.

Regional centers and centers for independent living

In some places like California, nonprofit entities called regional centers help coordinate support services for disabled people, including neurodivergent clients. They help people navigate the expansive web of services and information that they may not otherwise know about.

Similarly, across the United States, the Centers for Independent Living (CIL) are community-based organizations that provide a range of services and support to empower individuals with disabilities to live independently and participate fully in their communities. Between regional centers and the CIL, these organizations help individuals with referrals, assessments, understanding civil and legal rights, educational opportunities, housing support, equal opportunities, and accessibility. Check to see whether there are entities like the CIL and regional centers where you live.

Utilizing accommodations

All humans use accommodations. (We talk about this in Chapter 2.) Many neurodivergent people don't realize the range of accommodations that are available to them. This includes accommodation at work, in travel, and in many other aspects of life. If you've never asked for an accommodation, you may have some anxiety in doing so as it's unfamiliar to you. Many services and resources are available to help a neurodivergent person identify what accommodations they need and how to secure them.

Understanding your legal rights

Understanding your legal rights as a neurodivergent person can help you gain advocacy and support for necessary services, promote equal treatment and justice in legal situations, and protect you from abuse, discrimination, injustice, or harm. No one expects you to be a lawyer, that's why there are services that can help you out.

Navigating health care

Navigating the health care world can be quite the journey for someone who's neurodivergent. Sometimes health care providers are not clued in on the unique ways we communicate, handle pain, respect personal space, or even process things in our minds and senses. But here's the thing: Being open and candid about your neurodivergent traits, and giving some pointers on how folks can best connect with you, makes a major difference. John from our author team shares the following:

> As someone who's sensitive to noise, I always bring my noise-canceling head-phones to doctor's appointments. They help me tune out the background noises in the waiting room. If I forget them, my blood pressure reading tends to be sky-high initially. On those days, I make it a point to talk to the nurse about my neurodifference and ask if they could recheck my blood pressure later in the appointment. After spending time in the quiet exam room during my visit, my second reading always falls within the "normal" range.

REMEMBER

Just like your parents, your medical providers may not automatically grasp your neurodivergence. It's important to speak up and communicate your needs clearly to make sure you're well understood and supported.

5

Empowering Neurodistinct People

Discover universal perspectives and strategies that can help you be more effective at including, supporting, and empowering neurodivergent individuals in social, educational, and employment settings.

Recognize neurodivergent differences in cognition, perception, and communication, and how you can appreciate and support neurodivergent strengths.

Understand how to enrich your entire family dynamic when parenting a neurodivergent child.

Explore how neuroinclusive classrooms help all students, neurodivergent or not, put their best feet forward from elementary school through post-secondary education.

Discover why workplaces need neurodiversity and the importance of modernizing workplace practices and policies to help neurodivergent employees thrive.

Chapter **17**

Discovering Universal Perspectives and Helpful Strategies

D o you know someone in your life who thinks a little differently? Perhaps that's you, your child, a coworker, or even your neighbor. Or maybe you're a service provider or teacher who works closely with a client or student who has a unique mind. In this chapter, you explore some of the universal perspectives and strategies that can help you be more effective at including, supporting, and empowering neurodivergent individuals. Whether you're in a social, educational, or employment setting, these tips can help you make a positive impact on the lives of neurodivergent people.

Take Time to Understand Your Perspective

How you see and understand the world based on your experiences, beliefs, and thoughts is called *perspective*. It affects how you interpret events, how you feel about them, and how you interact with others. Some parts of your perspective are obvious to you, but there are hidden aspects that you may not be aware of. These hidden views still impact how you expect people to behave and the judgments you make. On our author team, Ranga grew up in India before he immigrated to the United States and often recognizes that cultural influence on how he views the world. He volunteers for a community help line where he regularly observes how trauma impacts the way some of the callers react to mundane everyday issues.

If you want to support someone who is neurodivergent, it's crucial to first understand your own perspective. Your way of seeing the world matters! Understanding your own viewpoint helps you better understand the perspective of others.

Your perspective isn't fixed and can change when you learn new things. For instance, suppose that you're afraid of dogs because you've heard stories of people getting bitten. But then you meet a dog trainer who explains that most dogs are friendly and well-behaved. They teach you about different breeds and safe ways to interact with dogs. This new information challenges your fear and shows you that not all dogs are dangerous. You realize that dogs can be loving companions. Learning can change your perspective and make you feel more comfortable around dogs.

It's okay if your current understanding of neurodiversity is limited. Our knowledge of it is relatively new, even though it has always existed. So, it's normal to not be fully informed. What matters is that you're open to expanding your perspective when you encounter new information.

Adopt a Broad Understanding of Neurodiversity

You don't need to know everything about a neurodivergent person to support them. Think about your best friend — you don't know everything about them, but they still see you as understanding and supportive. The same applies to neurodivergent individuals you know. Supporting them is like supporting anyone else in your life. Simply listen to them, show genuine interest, offer help and encouragement, and let them know they can rely on you.

Understanding that neurodiversity is normal

Neurodiversity has been a part of human life since the beginning, but people didn't always see it that way. Back in the day, when people showed traits that were different from the norm, others thought there was something odd, broken, or just plain weird about them.

But, here's the thing: Just because we don't understand something doesn't mean it's abnormal. Things can be perfectly normal even if we don't yet realize that they are. Electricity is a good example. Before we figured out how it worked, people had all sorts of wacky ideas about it (and none of them were correct). Some ancient Greeks even believed that static electricity came for the "souls" of fossilized tree sap. Other ancient cultures saw things like lightning and electric eels and thought they were connected to local gods or magic.

We now know better. Just as electricity has always been around, even when we didn't understand it, so has neurodiversity. It's nothing new. It's nothing bizarre. It's totally ordinary and normal. The only thing that has changed is how we understand and appreciate these important and ordinary parts of everyday life.

Understanding neurodiversity's role

When we talk about neurodiversity in this book, we're referring to the common wide range of differences in how the human brain functions. In the past, if your brain worked differently from what was considered "typical," it was often seen as a problem to be fixed. However, it is now understood that there is no single normal way for our brains to work. Everyone has their own distinct personality, interests, and talents because everyone's brains are wired uniquely. And that's not just okay — it's actually pretty amazing.

Imagine a world where everyone thought, spoke, and acted in the exact same way. It would be incredibly dull and boring, right? It's similar to how every song having the same melody would become monotonous, no matter how beautiful that melody is. Human brains, with their differences in thinking, information processing, and interaction with the world, bring depth, vibrancy, and innovation to society.

You may remember learning about bell curves in school. It's a chart that shows how things are typically distributed, such as favorite desserts, heights of people, or time spent commuting to work. In simple terms, the *normal distribution* in a bell curve is a way of displaying how things are spread out. But here's a curious question: "Which part of the normal distribution is not normal?" Well, that's a meaningless question because what makes the normal distribution normal is that it includes widely varying points from one end to the other. This natural diversity in human traits and characteristics, driven by our amazing brains, is perfectly normal! We should embrace and respect this variation as a perfectly normal part of being human.

Accepting the challenges people face

You may be thinking, "Does embracing neurodiversity as normal mean we ignore the challenges and needs of neurodiversity individuals who require a lot of support?" Nope. Quite the contrary. Embracing the normalcy of neurodiversity means recognizing and respecting the different ways the human brain can work and the different variations in how humans live their lives. And part of the recognition and respect is making sure that everyone who is neurodivergent receives the support they need.

Here's the scoop: Accepting neurodiversity as normal doesn't mean we brush aside difficulties or the fact that being neurodivergent can bring challenges. Conditions such as autism, attention-deficit/hyperactivity disorder (ADHD), and dyslexia often come with their own unique difficulties. And you know what? That's okay! Every human faces difficulties and challenges. Understanding neurodiversity as normal allows us to provide tailored help and support to the neurodivergent people in our lives.

So, don't worry! Embracing neurodiversity as a rather ordinary part of human life doesn't mean that we forget about support. Rather, it means that we appreciate and value the incredible diversity of the human brain while ensuring that each human receives the specific support that they need.

REMEMBER

The shift we're discussing here is all about our focus and where we put our efforts. Instead of trying to "fix" neurodivergent individuals, try to understand, support, and appreciate their unique ways of thinking and experiencing the world. By doing this, you can provide support that suits their specific needs much better.

As with most things in life, finding the right balance is key. On one side, it is important to make neurodiversity a normal and accepted part of society. This means being open-minded, respectful, and inclusive toward all individuals. On the other side, it is also important to acknowledge and address the specific needs of neurodivergent people. These two goals are not contradictory; they actually work together to create a society that values and embraces diverse ways of thinking. It's all about celebrating differences and ensuring that everyone receives the support they require.

Understanding how society is designed

Suppose that you're a right-handed person, like most people you know. Using scissors, can openers, and even shaking hands — it's all a breeze for you because everything is designed with right-handed folks in mind. Then one day, you meet a lefty. They tell you tales about how they struggle to use everyday items, or how writing in a spiral notebook is a real pain, or how they constantly bump elbows at dinner tables. So, what's your take on it? Is it their problem for being left-handed, or would you start questioning why the world is so skewed toward right-handers?

Here's a not-so-secret issue in our society: Many things are designed assuming that everyone thinks, learns, and behaves in more or less specific ways. This design works fine for a lot of people, but it unintentionally leaves out those who don't fit into those assumptions. That includes individuals who are neuro-divergent, such as those with autism, ADHD, dyslexia, or other neurodivergent conditions.

The trouble is that our societal systems, such as education, employment, and social norms, weren't really created with neurodivergent individuals in mind. They often follow a one-size-fits-all approach, which can pose challenges for those who have different ways of processing information, interacting with others, or learning.

REMEMBER

Society isn't purposely excluding anyone. It's just that these systems were set up based on the predominant, or neurotypical, way of functioning.

Here's the deal: We need to take a closer look at how our systems are designed and make some changes. By considering and accommodating the needs of neuro-divergent individuals, we can create a more inclusive and supportive society for everyone. It's time to broaden our perspective and make sure that no one is left out. Together, we can create a world that embraces and values the diversity of our brains! When we make society more accessible, it benefits all.

Ever seen those gentle slopes at the end of sidewalks called curb cuts? They were initially designed for wheelchair users, but guess what? They ended up benefiting more than just wheelchair users. Parents pushing strollers, individuals who use walkers, and even people lugging around heavy luggage found them helpful too. Pretty cool, right?

REMEMBER

If someone is facing difficulties, it doesn't necessarily mean they lack capability or aren't trying hard enough. It may simply mean that the systems they're navigating weren't designed with their body or their unique brain wiring in mind.

Realizing your community responsibility

No matter where you live, you are part of a community. A community is simply a group of people who have similar goals and help each other out. Communities can take many forms, such as the neighborhood where you live, online forums where people connect, or even religious groups of people who share the same beliefs.

While communities come in all shapes and sizes, there are universal traits which most communities share:

>> Every person deserves to be understood and respected.

>> Every person deserves compassion and empathy.

>> Every person deserves to be cared for, especially those who may be vulnerable or in need of support.

As community members, our role isn't to fix everyone around us. Instead, we should recognize the value of others, understand and respect their needs, show compassion and empathy, and provide assistance when needed. You can't single-handedly restructure society to support every neurodivergent person equally or fix all their challenges. But you can fulfill your responsibility within your community by caring for and understanding others, just as you would want to be cared for and understood.

Practice Strategies That Empower

To put these universal perspectives into practice and empower the neurodivergent people in your life, the first step is to educate yourself about neurodiversity, which you are doing by reading this book. Then, you can help educate others about neurodiversity and adopt practices that more fully support the lives of neurodivergent people. The following sections offer tips that can help.

Establishing your awareness

You've probably heard of phrases such as *autism awareness* or *dyslexia awareness*, but what does this "awareness" really mean? A simple analogy may help break it down.

Imagine being in a dark room with a huge elephant right in front of you. If you don't know it's there, you won't even notice. But when someone tells you about the elephant or you get closer to it, you become aware of its presence. You may smell, hear, or see it. The same goes for any issue.

Being aware means having knowledge and understanding. It's like shining a light on something previously hidden, allowing your brain to comprehend something and then take action.

Raising awareness in others

When you learn something important, you naturally want to share it with others, just like when you discover a cool gadget or a great movie. Raising awareness means getting more people to understand and care about the same issue as you. It's like forming a group of people who know what's going on and want to make a difference.

However, you can't reach everyone on your own. You have limited time and energy. But don't worry! Even small actions matter. When you talk to someone about the issue or share information on social media, you're contributing to what we call *herd awareness*. It's like herd immunity, when many people in a community are protected from a disease through vaccination. It makes it harder for the disease to spread to those who can't get vaccinated due to various reasons. The same principle applies here.

REMEMBER

Raising awareness about neurodiversity isn't just about focusing on the challenges faced by neurodivergent people, but also recognizing their humanity, strengths and perspectives.

TIP

Here are practical ways to raise awareness about neurodiversity:

>> Practice understanding and acceptance, and be patient, respectful, and accommodating.

>> Start conversations about neurodiversity with others to spread awareness.

>> Support organizations advocating for neurodiversity awareness by contributing or volunteering.

Taking these steps helps raise awareness, making a positive impact on the lives of neurodivergent individuals. Remember, every effort counts!

Exercising compassionate curiosity

It's easy to see why people think that their way of experiencing the world is the only way, but that's like saying vanilla is the only ice cream flavor. Just as there are many ice cream flavors, each person's brain is wired uniquely, creating diverse ways to experience the world.

Even if you *really* love vanilla ice cream, you may find new flavors that you like if you inquire about the preferences of others and are compassionate enough to understand their perspective. Compassionate curiosity helps us understand and appreciate others' experiences and preferences. It's about being open-minded and willing to learn. It means recognizing that there's no one-size-fits-all when it comes to thinking, feeling, learning, or interacting.

When encountering someone with a different way of thinking or acting, such as someone with autism, ADHD, or dyslexia, compassionate curiosity encourages understanding rather than assumptions and judgments. It promotes empathy, learning, and appreciating their unique perspective. By practicing compassionate curiosity, we create an inclusive world that celebrates diverse experiences and contributions.

TIP

Here are practical ways to practice compassionate curiosity:

>> Be aware of your own expectations and judgments; they don't help understanding.

>> Avoid assumptions about others; each person is unique and may not fit stereotypes.

>> Suspend judgment and seek understanding when someone's actions puzzle you.

>> Actively listen, ask questions, and empathize with the other person's experience.

>> Practice empathy, considering how neurodivergence impacts someone's daily life.

>> Be patient with yourself and others as you learn about neurodiversity.

>> Practice self-compassion; it's okay to make mistakes and learn from them.

By practicing compassionate curiosity, we understand and appreciate the diverse tapestry of neurodiversity. It enriches our perspective, enhances empathy, and builds a more inclusive society. It's like trying different ice cream flavors, each with its unique deliciousness. Why not create a world where understanding and acceptance thrive?

Accepting differences

Kids are taught that being different is special, but as kids grow up, they forget that lesson. Adults tend to expect everyone to be similar, which can be a problem for neurodivergent people. Neurodivergent individuals may approach tasks, communicate, or need support differently to succeed. We should remember the lesson we learned as kids: to accept differences in each other, including how our brains work.

And by "accepting differences," we mean more than just acknowledging that differences exist. It's about embracing and valuing the natural differences between individuals and understanding the importance and validity of these differences. There isn't a single correct way to have a brain. Believing that everyone must fit into the same mold is impractical and deprives the world of the richness that diversity brings.

Acceptance is being at peace with the reality of something, allowing you to shift your focus toward making a difference for the other person. Acceptance creates a sense of psychological safety for the other person, where they feel seen, heard, and understood. This allows them to be their authentic selves without masking their neurodivergence, empowering who they are to shine.

Here are practical ways to accept differences in neurodivergent individuals:

>> Communicate openly by asking about their needs and listening attentively.

>> Value different perspectives and appreciate the unique insights they bring.

>> Respect boundaries, considering comfort levels with social interaction and sensory stimuli.

>> Practice patience, allowing time for processing and expression.

Embracing the differences between all human beings enriches our lives and communities. When we accept and celebrate our diversity, we experience the world in all its wonderful glory.

Including for abilities

Every human faces challenges in life. But, when it comes to neurodiversity, people tend to *only* focus on challenges instead of a neurodivergent person's abilities and strengths. It's like only looking at the frames in an art museum, ignoring the beautiful masterpieces they hold. It is important to appreciate the whole picture, including the strengths of neurodivergent individuals. Accepting differences means embracing both the painting and the frame.

If someone is offering you a free gift, would you turn it down? Probably not! When we talk about "including for abilities" we're talking about accepting the valuable things that every person has to offer, whether it's a talent, a unique trait, or simply being there as family or friend. Unfortunately, the abilities and contributions of neurodivergent people are often overlooked. (To better understand strengths commonly associated with neurodivergent people, check out Chapter 18.)

REMEMBER

The abilities of neurodivergent people are as varied as those of neurotypical people. By approaching a person with compassionate curiosity, you can discover and appreciate their unique abilities, talents, and areas of strength.

TIP

Here are some practical ways to practice including people in social settings:

>> Be patient and open-minded, understanding that traditional social norms may not come naturally to neurodivergent individuals.

>> Improve communication by giving clear and detailed instructions and plans ahead of time and allow sufficient time for preparation.

>> Provide written information if preferred by the individuals.

>> Maintain consistency and predictability in routines for those who prefer structured environments.

Check out Chapter 20 for tips on how to better include neurodivergent people in educational settings as well as Chapter 21 for inclusion tips on workplace inclusion.

Chapter **18**

Understanding Differences and Embracing Strengths

Understanding how neurodivergent people think and communicate may seem tricky at first. But, we're here to help. In this chapter, you unlock neurodivergent differences in cognition, perception, and communication, and you explore how to appreciate and support neurodivergent strengths.

The term *cognition* refers to how humans think and process information, and it's essential for everything we do. It was once thought that all humans thought the same way, but it is now known that that isn't true. Neurodivergent folks think differently than most. Their thinking style works great for them, but it often presents challenges within a neurotypical world.

Neurodivergent thinking is totally normal, even if it's different than yours.

REMEMBER

Understanding Executive Functioning

An important part of cognition is *executive functioning*. In Chapter 5, we describe executive functioning as "how we get things done." It's how our brains help us accomplish tasks such as remembering, planning, reaching goals, and navigating obstacles. Due to different thinking styles, neurodivergent people often encounter difficulties with executive functioning in our neurotypical-centered world.

Here are some simple, practical ways you can help support a neurodivergent person when it comes to executive functioning:

>> **Organization:** Offer to assist with organizational tasks. This can be as simple as breaking a big task into smaller, manageable parts. It's like dividing a pizza into slices so that it's easier to eat.

>> **Schedules and routines:** Help them establish clear routines and schedules. It's like having a road map for the day, helping them know what to expect and when.

>> **Tools and technology:** Introduce them to task management tools that you find useful, such as digital calendars or note-taking apps. *Remember what works for you may not work for them!*

>> **Patience:** Give them time to process information and make decisions. Rushing things may overwhelm them or lead them to mistakes.

>> **Encouragement:** Celebrate their efforts and successes. This builds confidence and motivation. Keep praise genuine. Treat their victories the same as you would anyone else's.

>> **Guide, don't control:** Offer guidance and support, but let them retain control. It's their journey; you're just helping them navigate.

>> **Understanding:** Recognize that everyone has their strengths and weaknesses. Understanding and acknowledging these differences can help them thrive.

REMEMBER

You don't have to know everything about a person's neurodivergence to help them with executive functioning. Seek to understand them, treat their differences as normal, and offer support where you can. You'll see that can make a big difference.

Here's a practical example. Suppose you're talking with your coworker Paul, who's struggling to explain a project. If Paul's neurodivergent, factors such as bright lights or distractions may hinder his thinking. Try saying, "Hey Paul, what if we use the whiteboard to map out the project?" This small adjustment may be enough to help him focus, visualize his thoughts, and communicate the points he wants to make. See Chapter 21 for more about how support neurodivergent executive functioning in the workplace.

Embracing Communication Differences

Neurodivergent and neurotypical individuals often express their thoughts and feelings differently. Those unfamiliar may find neurodivergent communication styles puzzling, and they may struggle to interpret the intentions, facial expressions, or body language of neurodivergent people. It's crucial not to misinterpret these differences as deficiencies; they're just unique ways of communicating.

Consider this: A former student of ours, who often speaks in short, flat-toned sentences, got an exciting job offer. When asked whether he was thrilled, he whispered "yes" and quietly walked away. While some may misinterpret his response as sarcasm, we knew it was a mix of excitement, joy, and anxiety packed into that one-word answer.

WARNING

All humans communicate, even if some can't speak or have limited physical abilities. Don't assume they aren't communicating, as their methods are often misunderstood.

Here are some tips to help you engage neurodivergent communication styles:

>> **Flexibility:** Be adaptable in the ways you communicate. Some folks may prefer written communication over verbal, or vice versa. Find out what works best.

>> **Clarity:** Keep your language clear and to the point. Avoid using idioms, sarcasm, or complicated words that can lead to misunderstandings.

>> **Patience:** Allow them time to process what you've said and form their response. Think of it like slow-cooking — it may take a bit more time, but it's worth the wait.

>> **Rephrase:** If they're having trouble understanding something, try explaining it in a different way. It's like trying different keys until you find the one that opens the door.

>> **Active listening:** Show that you're paying attention and understanding their point of view. Nod, paraphrase, and ask clarifying questions. If someone talks a lot about a subject, it's fine to gently cut them off and say "Thank you for sharing that. I'd like to talk to you about something else now."

>> **Nonverbal cues:** Be aware of their style of nonverbal communication such as body language, facial expressions, and personal space. These cues can often tell you more than words.

Here's an example of some of these tips in practice: Autistic individuals or ADHDers can passionately talk at length about topics they are interested in. On our own author team, that describes John and Khushboo perfectly. Sometimes this gets in our own way when we're presenting to audiences. So, we use Ranga as an accommodation! He's really good about advancing us to the our next slide whenever we've accidentally gone over our time limit and he's found a natural breaking point.

REMEMBER

Everyone's communication style is unique. Talking with someone who communicates differently than you may require more time and patience. If it's difficult to understand the person at first, give it time and be open to learning their unique communication style. It's like learning a new language! Your efforts to support and respect those differences can help everyone feel heard and valued.

Decoding Social Behavior

The notion that neurodivergent people, particularly those with autism, don't like to socialize is a myth (see Chapter 6 for more about autistic traits and characteristics). Like anyone else, they crave community and connections. It's just that their way of socializing may not align with neurotypical norms, causing misunderstandings.

Neurodivergent individuals socialize in ways that sync well among themselves. But, understanding others' thoughts or actions from neurotypical nonverbal cues can be tricky for them. In neurotypical settings, this can lead to anxiety, over-thinking, or uncertainty about their inclusion. Guess what! Neurotypicals feel uncertain around neurodivergent people as well. So, neither group quite *gets* the other group. This means both groups need to have empathy for the other group and take active steps to overcome this gap. (See Chapter 6 for more.)

Here are some ways to make socializing with neurodivergent folks more inclusive:

>> **Talk it out:** Open a conversation about their preferences and needs. Remember, every person is different, so what works for one person may not work for another.

>> **Find common ground:** Rather than relying on small talk, many neurodivergent people connect well over shared interests or experiences. If you find a

hobby, topic, or experience that is shared with a neurodivergent person, chances are a deep connection may form.

>> **Preplan:** Give them a heads-up about what to expect in social situations, such as the number of people expected to be there, the noise level, or even the food being served.

>> **Create a comfort zone:** Make sure a place where people are gathering has at least one quiet space to retreat to if things get overwhelming, such as a cozy, safe corner.

>> **Be direct:** Because of differences in communication, not every neurodivergent person knows whether they're meant to be included just by reading context clues. *Tell them!*

>> **One at a time, please:** Try not to bombard them with too many questions or discussions at once. It's like trying to juggle too many balls at once — eventually, one is bound to drop.

>> **Patience, patience, patience:** Sometimes they may need a little extra time to process or respond. It's like loading a webpage — good things come to those who wait!

>> **Listen to understand:** Make a genuine effort to empathize and understand their perspective.

REMEMBER

Navigating social situations designed for neurotypical people can often cause anxiety for a neurodivergent individual. Neurotypical people may misinterpret neurodivergent speaking patterns, body language, or facial expressions as a lack of interest or empathy, all while the neurodivergent person may be deeply engaged or expressing empathy in their unique way.

TIP

If you observe someone struggling in a social setting, explore ways to include them. The goal is to make everyone feel included and comfortable. If you're neurotypical, remember to also speak up for your own needs. That way, everyone can be understood and supported.

Making Sense of Sensory Differences

Our senses are like our body's detectives, picking up clues about the world around us — everything from the smell of fresh cookies to the feel of a summer breeze. As with other aspects of neurodivergence, neurodivergent people may have different sensory experiences than others. They may pick up sounds that others may not notice, feel touch more intensely, or be more sensitive to how things taste. (To explore this more, check out Chapter 6.) Most closely associated with autism, sensory differences are also present in other neurodivergent conditions as well.

Each neurodivergent person's sensory experience is as unique as their fingerprints — what's unbearable to one may go unnoticed by another. Neurotypical folks often overlook these differences, causing confusion when a neurodivergent person reacts to certain sensory inputs. This can frustrate both parties, with neurodivergent individuals struggling to explain their experiences. But hey, you're here to understand better! By learning to help them navigate sensory differences, you're making a potentially huge positive impact on their lives.

Here's how you can help:

>> **Practice empathy and understanding.** Try to understand and respect a person's differences. A loud noise or bright light may not bother you, but for them, it may be overwhelming.

>> **Ask about preferences.** If you know someone has sensory differences, ask them about their comfort levels. This can be as simple as asking, "Is this too loud?" or "Is the room too bright?"

>> **Create comfortable environments.** Try to create environments that are sensory-friendly. This may mean lowering the volume of music, dimming lights, or removing strong-smelling objects.

>> **Be patient.** Sometimes a neurodivergent person may need more time to process information or may react unexpectedly to a sensory stimulus. Be patient and give them the space they need.

>> **Educate others.** Share this information with others, too!

It's helpful to think of sensory processing differences as neutral traits. Depending on place and context they can be a benefit or they can be a challenge. But, on their own they simple are what they are — one of the many differences we humans experience.

When sensory processing differences do become a challenge, it can be overwhelming and exhausting for the neurodivergent person to constantly filter and ignore a flood of sensory stimuli. It can make it difficult for the person to focus or think. At times, dealing with sensory input can be experienced as physical pain. Many tools and accommodations that can help a neurodivergent person deal with sensory input are available. See the appendix for some examples.

Just as you'd want someone to respect your preferences and needs, do the same for others. Every step you take toward understanding and accommodating those with different sensory experiences makes a difference.

Saying "Heck, Yes!" to Stimming

Ever found yourself doodling in a dull meeting or fidgeting when anxious? That's *stimming* (self-stimulation), a way our bodies handle strong sensations or emotions. Whether it's foot-tapping from stress or hair-twirling during a scary movie, it's all stimming. All humans do it, and neurodivergent folks often need to do it more, *which is perfectly okay.* Discouraging this natural action can cause discomfort and disrupt thinking.

Stimming is positive and should be supported. Here's how:

>> **Let it happen.** If it's not harmful to the individual or to another person, let the stimming occur. It's a perfectly natural behavior that is needed and is necessary in managing feelings and senses.

>> **Provide safe tools.** Fidget toys, stress balls, or even doodling can be helpful and allow for stimming in a safe, nondisruptive way. See the appendix for some more ideas.

>> **Educate others.** Let people know what stimming is, why it's important, *and that we all do it!*

>> **Don't shame.** Stimming can look a bit different, but that's okay. Never make someone feel bad for stimming and for doing what helps them cope.

>> **Encourage.** When someone stims in a safe and effective way, encourage and treat it as normal.

REMEMBER

All humans need to stim. This includes neurotypical people, too.

Appreciating Neurodivergent Strengths

Being neurodivergent simply means thinking and experiencing things differently, not being better or worse. We all have strengths and weaknesses (see Chapter 5 for more on neurodivergent strengths). Celebrating neurodiversity means embracing these differences. Historically, neurodivergent folks have been taught to mask their uniqueness, which can lead to feelings of shame and potential mental health issues. This can hinder them from living authentically and fully utilizing their strengths.

The good news is that attitudes are changing, and you can contribute to this positive shift. Here are some simple ways you can help:

>> **Value differences.** Life would be boring if we all thought and acted the same way. Let the neurodivergent person know you appreciate them for who they are.

>> **Play to strengths.** If they're great at spotting patterns, let them tackle tasks that need this skill. If they can focus on a single task for a long time, give them projects that need deep concentration. It's all about finding what they're good at and letting them shine.

>> **Patience is a virtue.** Neurodivergent people may take a bit longer to make decisions or may prefer to do things in a particular way. Remember, it's not about rushing, it's about getting it right. A little patience goes a long way.

>> **Comfort is key.** If they have specific preferences or needs, try to accommodate them. For example, if they find a quiet environment more comfortable, see whether you can minimize background noise. A comfy setting can make a world of difference.

>> **Be a cheerleader.** Encourage and celebrate their unique ways of thinking. A little praise can boost their confidence and help them realize how valuable their contributions are.

>> **Knowledge is power.** The more you understand their way of thinking, the better you can support and empower them.

REMEMBER

Don't judge a book by its cover! People often make assumptions about others based on what they see on the surface. They may think someone's disinterested because they're quiet or believe they're not engaged or intelligent. But wait! Instead of jumping to conclusions, try being curious and compassionate. As we first note in Chapter 11, by practicing compassionate curiosity, we create an inclusive world that celebrates diverse experiences.

IT'S OKAY TO LAUGH ABOUT OUR STRENGTHS

In her 2019 Netflix comedy special titled "Growing," actor and comedian Amy Schumer opens up about her relationship with her husband, chef Chris Fischer, and reveals that he was diagnosed with autism while they were dating:

> All of the characteristics that make it clear that he's on the spectrum are all of the reasons that I fell madly in love with him. That's the truth.

He says *whatever* is on his mind. He keeps it so real, you know? He doesn't care about social norms or what you expect him to say or do. Like, if I say to him "Does this look [bad]?" he'll go "Yeah. You have a lot of other clothes. Why don't you wear those?"

But, he can also make me feel more beautiful than anyone ever has my whole life . . . and he can't lie. *Is that the dream man*, a guy who can't lie? *But,* that also means that he can't lie *for me.* Which is an essential part of any relationship.

Schumer goes on to joke how her husband's refusal to tell white lies gets her in trouble among her friends. In doing so, she brilliantly portrays a mix of frustration, delight, and gratitude for her husband's neurodivergent traits. By sharing her joy and frustration, Schumer shows the audience that loving a neurodivergent person is no different than loving anyone else.

Chapter **19**

Caring for a Neurodivergent Child

When you have a neurodivergent child, the stakes aren't just high — they're different. Forget the cookie-cutter parenting manuals and the one-size-fits-all advice; your family is a living, breathing ecosystem of individual needs, and it demands a different set of nurturing tools.

This chapter isn't just about improving your child's life as their parent or caregiver; it's about enriching your entire family dynamic and deepening your own understanding of love, acceptance, and personal growth. In this chapter, you discover that the difference between *coping* and *thriving* lies in the nuances of support, understanding, and constructive communication. The happiness of your neurodivergent child can be a catalyst for a more fulfilling, harmonious existence for everyone in your household. You also discover that investing in your child's unique attributes isn't just beneficial — it's *essential*. And in doing so, you find that the very qualities that make your child different are the same qualities that can bring your family closer together than ever before.

This isn't just parenting; this is a paradigm shift. Welcome to a journey that will change not just your child's life, but yours as well.

Understanding Common Barriers

No one comes into the world feeling prepared for the journey of parenthood. The moment you hold that baby for the first time, or welcome a child into your home, you undoubtedly felt a surge of emotion and the sudden realization that you're not just responsible for yourself, *you're responsible for this whole other person!* It's not just about keeping them fed and putting clothes on their back; it's also about nurturing who they are meant to be. And that's a lot.

Those emotions you felt upon becoming a parent may well up again upon suspecting or learning that your child is neurodivergent. Before that point, you may have thought that you finally had your parenting road map all figured out. Then suddenly, what was printed on your road map may as well have been written in invisible ink. Poof . . . *gone!* For many parents, this knowledge can bring a head-spinning flood of anxiety as they try to get a grip on understanding what exactly this all means and where they go from here.

Have you ever had to unlearn everything you knew about something, only to have to rebuild your understanding all the way from the ground up? That's how a lot of parents describe their experience when raising a neurodivergent child. If you have felt the same, you are very much not alone. Successfully moving forward means realizing that the road map you have carried with you may not neatly apply from this point on. In fact, you may need to create a completely new road map of your own.

REMEMBER

All the parenting advice you may have been given can be helpful, but having you and your neurodivergent child thrive means figuring out the things that work best for your particular family and your particular needs.

The great thing is that this journey can be a beautiful experience. It doesn't have to be scary or challenging to create a new parenting road map that works best for you and your neurodivergent child. But, we get it. Navigating an unfamiliar terrain is often rocky and challenging — especially when you need to keep your balance between that journey and your obligations to other family members, work, and societal expectations.

However, we don't want to sugarcoat the barriers that may make this journey tougher than it needs to be. In the following sections we examine those barriers so that you can recognize and move around them.

If only you knew

Finding out that your child is neurodivergent often comes with a steep learning curve. You may have some surface-level knowledge of autism, attention-deficit/hyperactivity disorder (ADHD), or dyslexia, maybe from a TV show or a college psychology class. But suddenly, this isn't an abstract concept; this is *your child*.

The moment when you first learn your child's diagnosis is the best point of time for a health care provider to impart a deeper understanding of your child's neurodivergent condition, assure you that your neurodivergent child is still normal, and provide steps you can take to help them thrive (for a deeper understanding on the normalcy of neurodiversity, see Chapter 1).

Unfortunately, that doesn't always happen. So many parents are largely left to just figure things out on their own. We suspect that you may be one of these parents as well. Here are some common barriers parents face in getting the information they need. See whether any are familiar to you:

- » **Limited resources:** Many communities lack comprehensive resources on neurodiversity and neurodivergent conditions, leaving parents to sift through an ocean of information online, which can often be conflicting, misleading, or even full of scams and conspiracy.

- » **Outdated information:** Available information often doesn't reflect current scientific thinking and is still framed within a medical model, focusing on treating symptoms rather than embracing neurodiversity as a natural variation of the human experience (see Chapter 1).

- » **Stigma and stereotypes:** Society often stigmatizes neurodivergent conditions, leading to misinformation. Stereotypes perpetuated by the media can also distort a parent's understanding of their child's condition.

- » **Accessibility issues:** Scientific and medical literature can be hard to interpret for nonspecialists, and it may be locked behind paywalls, making valuable insights hard to reach.

- » **Conflicting advice:** Parents often receive different recommendations from health care providers, teachers, and other parents, leading to confusion about which strategies are truly effective.

- » **Cultural barriers:** Some cultures have differing views on neurodiversity, and those providing information to families may not fully understand the needs of specific cultures and communities. This can influence how information is provided, interpreted, or even whether it's sought out in the first place.

>> **Time constraints:** Parenting a neurodivergent child often demands a lot of time and attention. This can make it difficult for parents to find the time to educate themselves thoroughly on the subject.

>> **Financial barriers:** Assessments, adjustments, supports, and the energy needed to advocate for accommodations can be expensive, and not all families have the resources to access them. This may leave us with limited avenues to access accurate information.

>> **Lack of representation:** Most existing resources are centered around the experience of a particular demographic, often excluding minority populations, which makes the information less universally applicable.

Understanding these barriers can be the first step in overcoming them and empowering yourself and your child with the support you need.

REMEMBER

Having access to quality information about neurodiversity and your child's neu-rodivergent condition demystifies things, clarifies what to expect, and highlights available support. Understanding the barriers you may face in acquiring that information helps you move around them and gain the knowledge you need to not just cope but to thrive (for where you can start, see the appendix).

Take a moment and imagine a world where parents get a neurodiversity handbook right from the get-go. Upon their child being diagnosed with a neurodivergent condition, these parents are instantly connected to neurodivergent adults, health care professionals, and other parents who are also walking the same path. It would be like having your dream support network, right there when you need it most.

Imagine all the players — health care folks, insurance companies, schools, you name it — coordinate to help parents and their neurodivergent children thrive. That's the ideal but we're not quite there. So, we appreciate the responsibility that you bear right now in figuring much of this out on your own. We hope this book helps.

What you wish they knew

Your responsibility for advocating for your child's needs starts the moment you learn of their diagnosis. You may have walked into a health care office, but you walk out into a world filled with people with little understanding of the unique needs and strengths of your child. Few people get the ins and outs of your neuro-divergent child's experience like you do. So, you're not only navigating this new territory for yourself, but you're also suddenly having to act as a guide for every-one else as well.

While some people you know may become distant due to their own anxieties and fears, others out there can provide you with understanding and support. Perhaps they are other parents, experts, or neurodivergent adults. You're not alone. It may take a little work to find these folks, but it's worth putting in the effort to seek them out (for resources on how to find a neurodivergent community, see the appendix).

Lack of financial support

WARNING

When your child is diagnosed with a neurodivergent condition, you're often handed a list of people and providers who promise to make everything better for your neurodivergent child. Now, that list may have some supportive providers on it who can help you and your child thrive. Unfortunately, many resource lists also contain providers rooted in outdated ideas. Worse, those seeking to sell false promises while hoping to make a quick buck off your child's condition sometimes find their way on to those resource lists as well.

This means that right away you begin to spend a lot of time and energy pouring through these lists to find who and what may be right for your child. On top of that, many of the resources suggested to parents can be quite expensive.

No one is paying you to do this work. And most of us don't have the financial resources to pay others to do this work for us. However, many resources are available to assist you. This includes parents who've been there, done that. They can be a gold mine of information and can help guide you through what's actually worth your time and money. Ask your school, religious community, community group, or work colleagues for suggestions of people to connect with. Online forums, social media groups, and local organizations can help connect you with these parents as well.

TIP

Another great resource for finding appropriate support for your neurodivergent child is neurodivergent adults. Many of them grew up going through the same type of services and support you're examining now. Most major metropolitan regions have organizations and online communities of adults with particular neurodivergent conditions. We've found these groups to be regularly willing to take the time with parents who are trying to figure out a path that's best for their child. So, reach out! (For a starting point on where to connect with these groups, see the appendix.)

You should also know that you're not alone in facing financial pressures of securing the supports that your child needs. Look for grants, scholarships, or sliding-scale fees. And some organizations are specifically designed to financially support families with neurodivergent kids. For example, the iTaalk Autism Foundation helps educate families on the beneficial uses of interactive technology that enables children to communicate, and this organization has directly provided this technology to many families at no cost.

Limited availability of time and energy

Parenting in general takes a lot out of you. We know. You're juggling bills, work, household tasks, the needs of your neurodivergent child, the needs of other family members, and, you know, the *general chaos of life*.

Parenting is a monumental task, and you deserve so much credit for it. But remember, you're only human. The possibility of burnout isn't just real; it's practically banging on your door. You've got limited time and energy, and there are only so many plates you can keep spinning. It's okay to acknowledge that, really, it is. Because acknowledging it is the first step in getting the support you need to keep going.

We can only handle so much at once. Yet, the to-dos and the tasks never seem to end, and let's face it, you've turned multitasking into an Olympic sport. But guess what? people drop the ball sometimes — it's inevitable. Missed an appointment? Forgot a deadline? Took the wrong turn? It happens! And when it does, the most important thing is to cut yourself some slack.

That never-ending to-do list? It's easy to feel that everything is a *must do,* but keep in mind: Some things are *need* to do and some are *nice* to do. Knowing the difference can be a lifesaver when you're swimming in responsibilities.

And don't forget about the power of delegating or outsourcing tasks. Whether it's asking a friend to help organize your space, using a laundry service, or even getting an executive functioning coach to help your neurodivergent child, these external supports can make a world of difference. Not only does it help lighten your load, but it can also be a game changer for your child's growth and independence. Because when you have capacity, their capacity to do more things increases as well. How cool is that?

You're doing an incredible job, seriously. Perhaps you need to hear that because you may not feel that at times. You want people to extend your child compassion, patience, and grace, right? Sometimes it's important to pause and remember that you need to extend yourself that same compassion, patience, and grace as well. Have faith in yourself.

Coping as a Parent or Caregiver

Have you ever asked your parents how they coped while raising you? Most parents say that they just rolled with the punches when it came to parenting. There was no time for coping, just doing. But what about the daily grind? How did they handle the ups and downs that come with raising a kid? It's worth considering, because understanding how they dealt with the stress and incoming static can offer insights for your own parenting journey.

REMEMBER

Understanding your coping needs as a parent of a neurodivergent child is crucial. It helps you be a more present and effective parent. It's a key for resilience and stress reduction. When you're more resilient and less stressed, your child benefits too, feeling more supported and secure.

Unpacking coping

In simple terms, *coping* is how you handle stress, and humans generally go about it in two ways. The first way is called *adaptive coping*. This refers to employing strategies that prove to be healthy and effective. This is opposed to the second way of doing things — *maladaptive coping*. This refers to things you do to alleviate stress that may feel good in the moment (think smoking, procrastination, avoidance, or overeating), but often prove harmful in the long run.

TIP

Following are some examples of people using helpful (adaptive) coping strategies that may work for you:

>> **Problem-focused coping:** Sarah attends workshops on dyslexic-friendly parenting and implements structured routines for her child.

>> **Emotion-focused coping:** Mark practices mindfulness and deep-breathing exercises when he feels overwhelmed by parenting challenges.

>> **Social-support coping:** Emily joins an online support group for parents of neurodivergent children, finding solidarity and advice.

>> **Adaptive avoidance:** Lisa schedules "me time" to distance herself temporarily and recharge before tackling issues related to her child's care.

>> **Information-seeking:** Darnell reads books and consults experts on ADHD to better support his child's educational needs.

>> **Changing perspective:** Loida reframes challenges as opportunities for growth, turning moments of struggle into teaching points for her child.

Understanding yourself and your needs

Everyone has a breaking point — where it feels as though the weight of the world is just too much. If you've ever hit burnout, you know the time to ask for help isn't when you're at your wits' end. It's well before that — when you first notice things spiraling, or you realize you've been sidelining your own needs to support everyone else.

If you're parenting a neurodivergent child, you're so wrapped up in advocacy, in love, and in just the day-to-day grind that you may not even notice that you've been putting yourself last. And the thing is, it's easy to miss how long this pattern has been going on. So consider this your nudge to check in with yourself and maybe bring in some reinforcements before you hit that breaking point. Because you're not alone in this, even if it sometimes feels that way.

If this is your situation, it is important that you understand yourself and your needs. Recognizing your own needs is crucial, especially when you're responsible for another human being. That internal check-in — How am I doing? What do I need right now? — is vital. When you're stressed, the key is to catch negative emotions or thoughts before they spiral. Once you've caught them, pivot toward a coping strategy that actually helps you, rather than one that just pushes the problem down the road. Adaptive coping isn't just for the big moments, it's for the everyday stresses that add up. By consistently choosing healthier ways to cope, you're setting yourself up to be more present and effective not just for you, but for your child too.

REMEMBER

Coping strategies are only effective when practiced consistently. Start small and be patient when adopting new coping strategies. They take time to show results. Have you decided that exercising or meditation are right for you? Great! But, don't expect instant relief. It takes time, but the more you keep at it, the more relief you may find they provide. Consistency is key for effective long-term benefits.

Maintaining who you are as a person

When you're a parent or caregiver of a neurodivergent child, it's natural to prioritize the needs of your family over your own well-being. But remember that shifting part of your focus onto caring for your own needs isn't selfish; it's necessary for being the best parent you can be.

Allowing yourself a little bit of care

Pause for a second. When was the last time you thought about your own well-being? Your hobbies, relaxation, exercise, or even just grabbing a healthy meal that you enjoy? We know it's easy to put yourself last on the priority list when you're in the trenches of parenting. However, taking care of yourself is what sustains you to keep going on behalf of your family.

Finding pockets of time to unwind and recharge isn't just a nice-to-have, it's crucial. You're no good to anyone if you're running on empty. So make a promise to not lose sight of yourself while you're doing this incredible, challenging, and utterly important work. Your well-being matters, not just for you, but for your family as well.

Maintaining your own identity

If you're a parent or caregiver of a neurodivergent child, you already know the sheer amount of work and dedication it takes. It's a commitment that can easily consume your entire world, sidelining other important facets of your life — your hobbies, your social commitments, even quality time with your spouse or other family members. It can feel as though your individual identity gets subsumed by this incredibly important, but all-consuming, role.

You're more than just a parent of a neurodivergent child — you're still *you*. We've all seen t-shirts or bumper stickers that say "Autism Dad," "ADHD Mom," or "Proud Parent of a Dyslexic Student." And yes, dealing with the strengths and challenges of your child's neurodivergent condition is a huge part of who you are. But both you and your child still remain multi-dimensional people, with interests and qualities beyond neurodivergence. Make sure to nurture and celebrate those other parts of your identity; it's beneficial for you and sets a great example for your child.

REMEMBER

Maintaining your own identity is not just crucial for your well-being; it has a direct impact on your child's life too. Setting aside time for activities that fulfill your own needs and personal growth, you're teaching your child invaluable lessons about balancing life's complexities. It also enables you to approach challenges with greater resilience and a broader perspective.

Understanding your child and their needs

Understanding your neurodivergent child is crucial for their development and your family dynamics. This deeper understanding helps you better advocate for their needs, enhances your relationship with them, and fosters an environment where they can thrive. To understand your neurodivergent child, start by spending mindful time with them. Observe their behaviors, ask questions, and look for patterns. Even if they can't fully articulate their feelings or reactions, your attentive presence can offer valuable insights. Being curious helps you understand them better.

Understanding a neurodivergent child can be more complex than understanding a neurotypical one. Many neurodivergent kids can't easily express their needs in ways that neurotypical people understand, so observation is key. Signs of discomfort or disengagement indicate that something's not right, while increased engagement and smiles usually mean they're content. Their needs may vary — from needing time with a specific person to requiring a set routine. Teaching phrases, hand signals, or pointing to a picture to communicate "I like this" or "I don't like this" can clarify communication and enhance your interactions with them.

To truly understand your child, be an active observer and engage in their world. Ask questions, look for patterns, and observe how they respond to various situations. Quality time and a curious mindset can yield insights, even if your child can't articulate their feelings or experiences explicitly.

Discovering structures that work for your child

Understanding your child may include noting patterns of activity and behavior to determine a good daily routine for your child. Structured schedules with visual aids, timers, and clearly organized spaces can be a huge benefit to neurodivergent children, especially when they are given a bit of agency in defining them. When utilized with a mindset focused on a child's growth, they can open up opportunities for flexibility and adaptability to new situations as well.

In establishing structures, the best aids are the ones that work for you and your child. There are lots of apps, products, and approaches to consider. You may find friends or family members being really enthusiastic about recommending one of them to you — insisting that "it works wonders, and you just have to try it!" Try it. See whether it works for you, but it's okay if it doesn't. Every family is unique, and only you can determine what is best for yours.

Understanding your child's strengths

A neurodivergent condition isn't just a set of challenges; it's a unique lens through which the world is experienced, bringing with it distinct perspectives and strengths that our world not only benefits from but truly needs. Understanding your child's strengths can help you identify the kinds of learning styles and activities that are interesting and engaging to them.

How do you discover those strengths? Good question. Spend focused time with your kid. Whether it's building LEGO structures, solving math problems, playing video games, or cooking up a storm, you'll see where they excel. Keep an eye on what lights them up; that's usually a clue.

REMEMBER

Here's a little secret: Parents don't often think about how their kid plays with toys, looks at objects, or engages with media on their own. As long as they're not breaking anything, hurting anybody, or harming themselves, we tend to be happy as we juggle everything else. But how a neurodivergent child engages in those things can tell you a lot about their strengths.

Dyslexic kids may gravitate toward games that require strategy or seek out more visual or physical play. ADHD children may find things with varying textures (such as water, sand, or prickly pinecones) fun to play with, while autistic kids may really enjoy activities that allow them to use their observational skills. All of these things foreshadow associated strengths they may develop as they continue to grow.

And hey, consult with the professionals in their life — teachers, therapists, and doctors can offer unique insights about their strengths that you may not have. Bottom line: You've got a bunch of tools in your toolbox. Use them!

UNDERSTANDING NEURODIVERGENCE

We realize that we use terms such as *neurodivergent* and *neurodiversity* in this chapter. We discuss the meaning of those terms throughout this book (for a quick understanding, turn to Chapter 1). However, we want to acknowledge that — if you are a parent — you may have turned directly to this chapter without reading anything else. We're cool with that. But we also want to provide you with the simple definitions of these words in case they're helpful:

- **Neurodivergent:** Having a brain that functions differently from the majority of people. Studies suggest that this may be 20 percent of people or more.

- **Neurodiversity:** The range of differences in brain function and behavior that exists in the human population.

Our modern scientific thinking understands both neurodivergence and neurodiversity as normal. In fact, some neurodivergent conditions may have existed for as long as human beings have been around. We're just now beginning to understand how these differences benefit human society and how we can better support people who have these differences. Hey, we may be late to the party, but we're glad that we're here!

Balancing the needs of all children

If you have both neurodivergent and neurotypical kids, balance is key. Kids notice if one sibling gets more attention. Accommodating everyone's needs — including everyone's need for quality time and attention — benefits the whole family.

FOSTERING NEURODIVERGENT JOY

Jen White-Johnson, an artist and advocate, coined the phrase *neurodivergent joy* to highlight the unique happiness and fulfillment that neurodivergent individuals find in their own ways of experiencing the world. Unlike traditional discussions about neurodiversity that may focus more on challenges and accommodations, neurodivergent joy is about celebrating the distinct perspectives, experiences, and skills that come with being neurodivergent.

Inspired by the joy she saw reflected in her autistic son Knox, this amazing mom took her observations and channeled it into a powerful movement using her design skills. White-Johnson began to create stickers, posters, and other forms of media centered around neurodivergent joy (see the following figure). This resonated with neurodivergent adults, who'd often only heard negative perspectives on their conditions. It was a game changer. These adults started sharing their own joys and connected with parents of neurodivergent kids, spreading positivity and a fresh perspective on what it means to be neurodivergent.

Image courtesy Jen White-Johnson

Fostering joy within your child regarding their neurodivergent experience is essential for their well-being, personal growth, and mental health. Your kid is going to have some challenges in life, but they are also going to have a lot of joy in being able to experience the world in ways that many others can't. Celebrate that! And you know what? Doing this doesn't only help your child, it's good for you too. When you focus on what aspects of their neurodivergent experience makes them genuinely happy, it can change the entire dynamic of your family, making life more enriching and yes, joyful, for everyone involved. To access White-Johnson's resources, see the appendix.

While meeting the unique needs of your neurodivergent child is important, don't underestimate the universal benefits of good practices like a stable routine and clear communication for everyone. It's not just your neurodivergent child who thrives with predictability and understanding; your neurotypical kids do too. Making these adjustments is like upgrading the whole family operating system. Everyone feels more secure, more understood, and more included.

Mapping and Securing Your Support

We've all heard the saying "no man is an island." That's especially true for a parent of a neurodivergent child. We are all interconnected in so many ways, and one of the most productive things you can do is to map out and secure your support systems.

You see, for parents and caregivers of neurodivergent children, building a robust support system is vital for both your well-being and that of your child. Start by identifying family members, friends, and professionals who are knowledgeable about neurodiversity and can offer emotional or practical support. Connect with educational and health care teams who are experienced in meeting neurodivergent needs. Don't underestimate the power of community — seek out local or online support groups, workshops, and seminars that focus on your child's specific condition.

Building a robust support network isn't just a nice-to-have, it's a need-to-have. You're going want family in your corner, babysitters who get neurodiversity, and solid connections with teachers and health care professionals. If your family is part of a religious group or a community organization, it's helpful to identify people who are willing to understand and assist you there too. Don't forget the everyday stuff, such as carpool coordination and workplace allies who understand your family's needs. All these pieces together create a support web, making everyone's life a little bit easier.

These networks not only provide emotional backing but can also be excellent resources for sharing strategies, advice, and information. The aim is to create a network of support that you can rely on in different scenarios, making the journey more manageable and enjoyable for everyone involved.

You may be surprised to find out that some of your best support team members can be your coworkers. John from our author team tells the story of once speaking at the American headquarters of the French retailer Sephora:

> I wasn't there to speak on neurodiversity, but in my remarks, I happened to mention that I was autistic. During the question-and-answer portion of the program, an executive stood up and said, "I have an autistic child. I've never met an adult with autism before, so I really appreciate you being here." Suddenly, another executive in the back of the room jumped up and shouted, "I have an autistic kid as well. I thought I was the only one!"

After the session, John connected the two executives with each other. Many parents think that they're the only ones in the office who are raising a neurodivergent child, but that's often not true. In addition, most organizations have neurodivergent employees as well. Linking up with coworkers who just "get" you can help you push for the accommodations and flexibility you need as a parent at work.

Advocating for Your Child's Needs

Until your neurodivergent child can advocate for themselves, you're pulling double duty — fighting for both your own needs and theirs. To be effective in this role, it's crucial to get a deep understanding of what your child and your family truly need. Needs aren't always self-evident; they may take a little reflection to figure them out. They're nothing to be ashamed of. Being able to recognize and meet your needs is essential to being able to thrive in life.

Understanding the needs of your family

To effectively advocate for your neurodivergent child, the first step is to see their traits as a normal part of who they are. Understanding and accepting your child's unique strengths and traits is not ignoring or glossing over their challenges. Rather, it's about approaching those challenges from a place of acceptance and strength, which allows for a more constructive and empowering advocacy. When you see your child's needs as normal, you learn to recognize barriers and systems that may be preventing them from being treated just like any other kid.

As you reflect on the needs of your family, document your child's strengths, triggers, sensory issues, favorite learning methods, along with their preferred social and communication styles. Pay attention to what they need to do differently in order to get things done. As a parent, reflect on where you spend your energy, what exhausts you, and what contributes to your energy and resilience. Examine those things in your child too. Being familiar with all these things helps you more effectively guide teachers, health care providers, and service professionals as they help you.

Understanding your legal rights

Understanding your legal rights and the applicable laws can guide you to make informed decisions for your child's well-being. Different locations have varying laws on disability support services. Being aware of these regulations helps you know what you're eligible for and how to access and maintain those services. In addition, confidently knowing that the law is on your side in certain places can be a huge boost as you advocate for your family.

Advocating for support services

Navigating life as a neurodivergent individual can often mean tapping into various support services, such as occupational therapy or educational coaching. Lucky for us, there's a spectrum of funding options available from government programs to nonprofit initiatives. Do your homework to figure out what you're eligible for and how to apply, because let's be real; these services are in high demand and the costs can add up if you're doing it alone. So, get on those waitlists and take those first steps.

Advocating in the classroom

A major role you play as a parent is navigating advocacy in education. This includes making sure your child is in a classroom that works for them. It's also about ensuring that they have the support and accommodations they need and that the educational goals set for them are actually going to be helpful for their growth and happiness. We know it's a lot, but it's crucial.

TIP

Other parents have been on this journey, and many have battle scars to prove it. We've found that parents who have success in advocating around educational obstacles are happy and eager to pass the tips of their knowledge along. Talking with these parents and learning from their experiences may make the experience much smoother for you.

It can be a mixed bag when it comes to how well teachers are prepared to support neurodivergent students. Some may be experts, others may have taken a seminar or two, while still more may have never received any kind of specialized training. This gap can affect everything: the resources available, how well teachers adapt their teaching style, and even how willing they are to make changes. As a parent, you may need to fill in those gaps in knowledge by advocating for your child and the support they need.

TIP

Advocating doesn't necessarily mean *adversarial.* Being able to collaborate with teachers and school administrators in a meaningful way can positively benefit everyone involved. Your child gets the support they need, and your school grows their knowledge of neurodiversity and student needs.

REMEMBER

Here's something fun: Making classrooms more inclusive isn't just a win for neurodivergent kids. It levels the playing field for everyone — whether they're bilingual, from diverse backgrounds, or have other specific needs. Inclusive education is good for all, not just some.

Ensuring classroom integration

In the United States, students are either placed in a special education classroom with other students with disabilities or mainstreamed into a regular classroom with neurotypical peers. Classroom integration is dependent on the student's ability to keep up with the course requirements and actively and meaningfully engage with the teacher and with classroom peers.

When it comes to educating neurodivergent children, you child may end up in one of several different places. Some may be in what many still term *special education* classes, but really, these are specialized support environments designed to meet specific needs. Other kids may find themselves in a mainstream classroom but with some added support — think of it like custom-built scaffolding for their education that's designed around how they learn. And then you have the kids who are in mainstream classrooms with zero support, no accommodations, and frankly, no clue on the school's part about who they are or how they learn.

In recent years, the educational landscape has been changing — slowly, but it's happening. There's a growing emphasis on inclusion, aiming to integrate neurodivergent kids with diverse learning needs right into mainstream classrooms rather than segregating them. Sounds great, right? Well, it is, but this shift comes with its own set of challenges to navigate.

From skeptical teachers to bureaucratic red tape, advocating for your child's integration may require some persistent effort on your part. Even schools with the best intentions can fall short. They may not be equipped with the resources or, let's face it, the know-how to make these integrated classrooms work for every student. What can you do?

>> **Familiarize yourself with your child's rights under the law.** Knowledge is power, after all.

>> **Build relationships with the educators and administrators.** People are more willing to work with you if they know you.

>> **Don't underestimate the power of a polite yet firm email to keep things moving.** It's not always an easy road, but your advocacy makes all the difference.

Securing classroom accommodations

The word *accommodation* can scare a lot of people, but remember: Accommodations are just tools for leveling the playing field. If your child needs specific classroom accommodations, that's not just okay, it's actually good. Advocating for what your child needs is not special treatment; it's just making sure they have the same shot at learning and thriving as everyone else (for more on the normalcy of accommodations, check out Chapter 2).

Common accommodations in the classroom include extended time for tests or projects, visual learning tools, audiobooks, sensory-friendly spaces, the ability to move when needed, allowing oral responses instead of written ones, and the use of assistive technology.

REMEMBER

You're the advocate here — the one who's got to open up that line of communication with the teachers and the school staff. Be proactive! You want to arm yourself with knowledge. Know the laws, understand the school's policies, and most important, know your child's rights.

TIP

If you're wondering where to start, don't worry. Many organizations, government offices, and resources can assist you along the way (for connections to these, see the appendix).

Utilizing IEP plans at school

Something many families with neurodivergent children utilize in the classroom is an individualized education plan (IEP). Now, if you're not in the United States, you may know this by another name, such as a *Statement of Special Educational Needs*. The idea is the same: It's a legally binding document that lays out exactly what a school is obligated to provide for your child. We're talking special services, accommodations, you name it, to make sure your kid gets the most out of their education.

How do you use it? First, get involved in the IEP meetings. This is where you, the teachers, and usually a bunch of other professionals sit down and figure out what your child needs to succeed. You're not just a spectator; you're a coauthor of this document! And remember, an IEP isn't set in stone. As your child grows and changes, the plan can too. Keep those lines of communication open with the school, and don't hesitate to advocate for what your child needs. This document is more than just paper; it's a powerful tool in your advocacy toolbox.

An IEP meeting typically comes about in a couple of ways. First, a teacher, school counselor, or other education professional may notice that a student is struggling academically, socially, or emotionally and recommend an evaluation for special education services. This usually triggers a process that involves assessments and eventually leads to an IEP meeting if the student is found to be eligible for special services.

Parents can also initiate this process. If you notice that your child is struggling and feel that they may benefit from specialized instruction or accommodations, you can request an evaluation yourself. You usually need to do this in writing and send it to the school's special education department or the principal. After receiving the request, the school typically conducts an evaluation to determine the student's eligibility for an IEP.

REMEMBER

Keep in mind that parents often face challenges in the execution of an IEP, such as lack of clear communication with the school, inconsistencies in how accommodations are provided, or even resistance from teachers who may not fully understand the child's needs. In addition, the paperwork and bureaucracy involved can be overwhelming. Despite these hurdles, an IEP can be a powerful tool for parents to ensure their child's educational needs are met.

TIP

To successfully use an IEP, parents should maintain open lines of communication with the school, regularly monitor the implementation of accommodations, and not hesitate to advocate for adjustments as needed.

You should also remember that you have a right to question which goals are being targeted, how they are being run, as well as whether the measure of progress is valid for the student's long-term goals in regard to independence. If necessary, you may want to seek out an IEP advocate or an IEP lawyer to help you navigate this process, understand the contents of the IEP, as well as walk away with clarity on next steps. Being proactive and involved can go a long way in making sure your child thrives in their learning environment.

Aiding academic progress and success

Academic success isn't a one-size-fits-all situation, especially for neurodivergent students. Some subjects may come easy, while others feel like a mountain to climb. The way your child absorbs and communicates information may not fit the traditional mold. If they're in a system that doesn't get that, focusing in class or acing tests can become a serious challenge. And let's be clear, that's not on them; it's a sign that the educational system needs to adapt. A lot of neurodivergent folks state that the usual academic playbook didn't do them any favors. It's vital to recognize that different doesn't mean deficient.

Supporting socialization

A neurodivergent child may socialize differently than others. This isn't a bad thing. Studies have shown that many neurodivergent people socialize in ways that are immediately familiar and comfortable to others who share their neurotype. Just because how someone socializes or communicates is unfamiliar to you, it doesn't automatically mean that it's bad.

WARNING

It used to be standard to teach neurodivergent kids to suppress how they naturally communicate and socialize in order mirror neurotypical behavior. We now know this as incredibly harmful, with neurodivergent adults now sharing the trauma they accumulated by going through this process as kids. Unfortunately, a huge number of providers still rely on this old model even though our understanding of neurodivergent communication and socialization has evolved. When seeking outside assistance in helping your child socialize, look for providers who provide assistance from a neurodiversity frame (see the appendix for more).

Helping your child understand and accept their neurodivergent identity is crucial for them. Teach them that just as they have their own ways of socializing, so do others. Highlight how different social styles benefit different people. This lays the groundwork for building stronger connections between your child and the world around them.

John on our author team compares this to learning a second language. You don't want your child to change. Encourage them to communicate and socialize in ways that are familiar to them. But understanding another "social language" can be beneficial. The reverse is true too: Teaching neurotypical kids to socialize in ways familiar to your child is equally helpful.

TIP

Group activities can be amazing tools for learning and growth. By setting up flexible groups based on shared interests or skill sets, you can help your child interact with a diverse range of peers. They get to lean into each other's strengths and differences, creating this enriching environment where everyone grows together. It's like turning your living room or backyard into a mini think-tank, where kids not only get to have fun, but they get some real-world skills as well. These skills include teamwork, problem-solving, communication, and decision-making.

Addressing bullying and harassment

Dealing with bullying and harassment is an unfortunate reality for many neurodivergent kids, especially in school settings. The way they interact can sometimes make them targets, which is just not cool. Parents, it's key to equip your kids with the tools to identify and handle these situations. Role-play videos can be useful, helping them quickly discern between friendly and not-so-friendly interactions.

TIP

Sit down with your child, talk through hypothetical scenarios, and brainstorm coping strategies. Problem-solving these situations beforehand prepares them to handle real-world moments.

But hey, remember that this isn't just a neurodivergent issue. Establishing clear expectations for behavior across the board helps create a community where everyone knows the rules of the game. This can go a long way in cutting down bullying and making sure everyone gets a fair shot.

Advocating for medical needs

When advocating for the medical needs of your neurodivergent child, it's crucial to be proactive. Be sure to share your child's specific needs and sensitivities with the medical staff to help them understand your child's uniqueness. If your child experiences pain differently, doctors need that information. Using visual aids such as an emoji pain chart can be incredibly helpful. For non-speaking children, having assistive communication tools readily available is transformative in a health care setting. Being well-prepared ensures a smoother experience for everyone.

Planning for the Future

Planning for the future can be daunting, especially for a parent of a neurodivergent child. Fears about what may happen when they're out on their own or when you're not around to advocate for them can be overwhelming. The best preparation is teaching your child to be as independent as possible. Focus on instilling skills like self-determination, self-compassion, and self-advocacy. This sets them up to navigate challenges, speak up for their needs, and ask for accommodations.

TIP

While many neurodivergent kids become independent adults, some may need lifelong support. It varies from case to case. Just know that you're not alone; many have walked this path before you. It's worth searching them out as they'll have valuable insights for you and your child.

Planning for advanced education

Not every child's path leads to college, and that's perfectly okay. Vocational schools, trade programs, and apprenticeship programs are often a great fit for many neurodivergent learning styles. Every human being has an inherent need for growth and acquiring new knowledge, regardless of the form it takes. Planning for your child's future education is crucial, whether that involves university, trade school, or on-the-job training. Even if your child may not be able to work in a traditional sense, continued education still plays a pivotal role in enriching their life.

Education can take many forms — it can be specialized therapy to improve life skills, community programs that offer social engagement, or simply pursuing hobbies and interests that make life fulfilling. The key is to focus on facilitating a learning environment that aligns with your child's unique strengths and challenges. Doing so not only fosters personal growth but also equips them with the tools they need to navigate life as independently as possible and enriches their sense of purpose and self-worth.

For those seeking college education, there's a lot to figure out — where the school is, what kind of living situation your kid will be comfortable with, how heavy the course load is, how the school provides instruction, and of course the cost. Educational institutions that readily offer accommodations and have a disability office are generally better able to support neurodivergent learning needs than ones that require every student to fit a particular mold of learning.

Planning for employment

Planning for your neurodivergent child's future employment is a necessity. A job isn't just a paycheck, it's a source of structure, self-esteem, and purpose. And let's face it; the job market isn't always forgiving, especially for neurodivergent individuals. That means it's crucial to help your child identify their strengths and interests early on. With the right mix of encouragement, exposure to different career paths, and some hands-on training, you can help set your child up for a more secure and fulfilling adult life.

REMEMBER

Your role is to not get your child a job, but to help your child figure out how to find work and maintain a career themselves. That doesn't mean that they may not need support. There's a lot of barriers out there that prevent many neurodivergent people from finding employment. Fortunately, you can seek help from organizations and government entities dedicated to helping reduce these barriers for neurodivergent people seeking work (for how to connect with such entities, see the appendix).

Planning for independent living

When you're raising a child, you're also shaping a future adult, so it's important to equip them with the skills and knowledge they'll need for that future. *Independent living* refers to an individual's ability to manage their life with as much autonomy and choice as possible. It's about making your own decisions, from choosing where to live to how to spend your time. For people with disabilities, independent living may involve various levels of support, such as in-home care or assistive technology, designed to enhance autonomy.

Independence is tricky, especially for neurodivergent folks. When your child is young and at home, it's easy to keep everything running smoothly for them. But as they grow, it's not just about providing support — it's about teaching them to take the reins themselves. That's where independent living skills come in. Teach them these essential skills, let them practice, and sometimes fumble or fall. Be there to reinforce their growth and provide a safety net for their practice. Keep them steady but step back where you can, all the while ensuring that your love and support is still there.

Teaching independent living skills to younger kids

While many skills crucial for independent living are often assumed to be picked up naturally as kids grow, neurodivergent children may benefit from more explicit instruction. Teaching essential life skills such as potty-training, toothbrushing, dressing, preparing simple meals, doing chores, or keeping track of personal items can start early in childhood. These skills may be affected by sensory sensitivities, differences with motor skills, as well as memory. Chaining several steps of these skills to be taught in sequence, using visual reminders, keeping items in designated locations, and actively reinforcing your neurodivergent child's independence can help improve progress in all areas of daily living skills.

TIP

Involving your child in grocery shopping can be a lesson in budgeting, while cooking together can teach them about meal preparation and safety. Because many neurodivergent kids benefit from specific teaching methods, consider using resources such as visual aids, step-by-step guides, or reminders on their tablet or phone to reinforce these skills. This explicit focus on teaching life skills can help prepare your neurodivergent child for the complexities of adult life, enhancing their ability to live as independently as possible.

Teaching independent living skills to teens

As your child grows into a teenager, consciously introduce additional living skills such as money management, cooking, cleaning, organization, basic household or automotive repairs, emergency preparedness, exercise, and time management.

Don't forget skills related to social interactions, community involvement, conflict resolution, and future planning as well.

Planning for lifelong care

Some neurodivergent individuals need lifelong care. Planning for this is a key responsibility of their parents. Think about backup plans if you're unable to provide the care yourself. Options can include in-home support, shared homes that provide support yet still foster personal autonomy, or other supports tailored to your child's specific needs.

TIP

These options can be difficult to navigate. Few find it easy to do it alone. Connect with other families that have experience here, along with organizations dedicated to helping people live independent lives (see the appendix for more).

REMEMBER

Independent living practices also apply to a person needing support as an adult. From a disability perspective, independent living doesn't necessarily mean living alone or without any assistance. Rather, it refers to having control over one's life and the freedom to decide where to live, how to spend one's day, and who to interact with, among other things. It also means having access to resources and support systems that enable such choices and control.

Planning for financial security

It's time to get serious about something that you may have pushed off thinking about: planning for your neurodivergent child's financial future. Beyond basic things like setting up a savings account or a trust, you've got to look at the big picture. Where does your child see themselves in the future? What are their goals and dreams? Understanding their aspirations helps you and your child (yes, this should be a participatory thing for you both wherever possible) take more informed, meaningful approach to not only securing their financial stability, but also a future that resonates with who they are.

After your family has a grasp on your child's aspirations, it's crucial to think through the practical steps that'll get them there. Will your child live independently, or will they need lifelong support? Understanding these variables enables you to tailor your financial planning in a way that genuinely supports their envisioned future. We're not just talking numbers here; we're talking about shaping a life that aligns with who your child is and where they hope to go.

REMEMBER

This isn't just for families with substantial financial resources. Every family, regardless of their current financial standing, should be engaging in this kind of planning. Adopting financial strategies and habits is universal. In addition, there are options and support for all economic levels, including government benefits to community programs, that can contribute to your child's long-term financial stability.

Helping your child establish financial habits

As your child transitions into adulthood, start identifying your child's strengths and challenges in the realm of financial planning. Use this knowledge to customize learning experiences and to select the most suitable tools and resources that fit how they learn. Encourage independence but remain available for guidance, providing a safety net as your child begins to manage their own finances.

TIP

For some neurodivergent individuals, cognitive challenges may make tasks such as budgeting and personal financial planning a bit more complex. But complexity doesn't mean impossibility — it often just means that a tailored approach is needed. Help can come in various forms, such as personalized financial coaching, which can break down complex topics into manageable steps. Consider assistive technology such as budgeting apps with user-friendly interfaces that offer visual aids and reminders to simplify the planning. Establishing and practicing effective habits, perhaps through structured routines or checklists, can further facilitate financial management skills.

Using financial planning resources

Financial planning resources include tools, information, supports, and services designed to help you navigate the world of money and investment. Think budgeting apps, investment platforms, how-to books, educational workshops, savings calculators, or even experts like financial advisors.

For parents of neurodivergent kids, these tools can be especially helpful due to the unique needs you may have as you create a financial road map for your child's adult life. Whether you're considering long-term care, educational paths, or helping your child establish an independent adult life, these resources can help you figure out what benefits are available to you, how much to put into a savings account or trust, where to invest your current resources so that they can grow, or how to set up a sustainable budget that factors in your family's unique needs.

TIP

Be proactive in seeking out financial planning resources, be they specialized financial education programs for neurodivergent individuals, planning workshops, or connecting with professionals who have relevant expertise.

It may be beneficial to seek social services and benefits specifically designed for neurodivergent individuals. Grants, scholarships, and employment programs can offer substantial financial support. Also, consider consulting a financial advisor who specializes in disability planning; they can help you make sense of the complex financial landscape. Planning for financial security is not just a box to tick off; it's an ongoing process to ensure that your child can lead a fulfilling life in the long term.

REMEMBER

Most important, approach financial planning as a collaborative effort with the presumption of competence; you and your child are on this journey together, and each step you take is one toward empowering your child for a more secure and independent life. You can never know for certain how far your child will reach for their dreams!

Chapter **20**

Supporting Neurodivergent Students

N eurodivergent minds are present in every classroom, each with their unique way of learning and thinking. But hold on; a lot of them aren't getting the support they need. The problem? Many schools and lessons just aren't tailored for neurodiversity. No worries, though! In this chapter, we dive into simple strategies to back our teachers, update our education methods, and cook up neuroinclusive classrooms that can help all students put their best foot forward.

Valuing Neurodiversity in the Classroom

As a teacher, neurodiversity may be a new term for you, or perhaps it's an everyday part of your teaching life. At the least, most teachers have come across phrases such as *special needs* or *special education* even if they're not part of your own curriculum.

It may be that you've never thought that you may have students with autism, attention-deficit/hyperactivity disorder (ADHD), dyscalculia, dyslexia, or other neurotypes in your classroom. And that's totally understandable. Just as neurodivergent students and their families largely have not been given the information and resources that *they* need, pause for a moment and recognize that *you* may not have been handed the tools, information, or resources that *you* need to fully assist neurodivergent students. Sound familiar? It's not your fault, and you're far from alone in this journey.

To all educators out there, whether you work full-time with neurodivergent students or are just beginning to explore what neurodiversity means, we shed light on this topic, step by simple step, in the following sections to give you the tools you need to help neurodivergent students thrive. The ultimate goal is to make the classroom a welcoming place for every mind, and you're just the person to do it.

Roughly one in five individuals are neurodivergent. Many may not entirely understand their differences' impact on their learning and may hide their struggles.

Neurodivergence isn't always obvious, so your classroom probably has neurodivergent students. Learning to incorporate neurodiversity in your classroom offers many benefits. One of these benefits is that the approaches that help neurodivergent students tend to help other students as well. This is particularly helpful for bilingual students, students from cultural backgrounds different than the majority of peers, and students who have transferred into a new school. All of these students, like their neurodivergent peers, are navigating an educational track that is not designed to meet their specific needs.

Understanding Barriers Educators Face

Neurodivergent students aren't the only ones who face challenges in the classroom. Teachers do as well. Have you been offered the training and support you need as an educator to successfully teach neurodivergent learners? Probably not. In fact, multiple barriers are routinely thrown in front of educators. We walk through a few of them in the following sections and discuss what you can do about them.

Lack of information

The first barrier educators often face is a lack of information on neurodiversity and neurodivergent conditions. And this lack of information is not simply an academic hurdle; it's a personal and professional roadblock. Without understanding how neurodiverse students think and learn, teachers can find themselves in unfamiliar territory, struggling to make those vital connections. This isn't just about teachers feeling lost or unsupported; it's about grappling with feelings of inadequacy or failure that can choke off the creativity and passion that are central to great teaching. It's a complex issue, but one that underscores the profound importance of understanding neurodiversity in the educational landscape.

The challenges caused by this lack of understanding go beyond student engagement; they strain the very heart of what it means to be an educator. By taking the time to learn about neurodiversity, teachers can begin to break down barriers in the classroom and connect with the passion that got them into teaching in the first place.

By now, most educational institutions require teachers to gain at least a basic knowledge of how to support students with disabilities. However, almost no school district in the United States provides training on how teachers can understand and support neurodivergent learners. The level of understanding and experience that any educator has in regard to neurodiversity varies widely and largely relies on whatever information and resources the educator has been able to cobble together themselves. *How exhausting!*

If you feel behind the curve in understanding neurodiversity, take a breather. Almost every teacher is. Just by reading this chapter, you've taken an important step toward positioning yourself as a leader in helping our schools better support neurodivergent minds.

In today's classrooms, teachers need to know how to mix things up. They need the right hands-on practice and ongoing training in how to adapt the ways they teach.

Lack of support

Even with the best training, educators may still find themselves in a tough spot. Why? Because they may not have the necessary backing from their faculty or leadership to consistently support neuroinclusive classrooms. It's like having all the ingredients but not enough pots and pans. Teachers often lack crucial resources, such as support staff, special education specialists, and the right instructional materials.

You know, it's not exactly a secret that many teachers today dig into their own pockets to buy basic classroom materials. So, it's a good bet that our educators aren't being provided with the guidance and materials they need to help neurodivergent minds thrive. If teachers want to foster a classroom that engages and uplifts neurodiversity, they need more than just dedication and good intentions to support diverse learners; they need real tools. School leadership needs to provide neurodiversity training and adaptable classroom materials that cater to all kinds of minds. It's not just about wanting to reach everyone; it's about having the means to actually do it.

Lack of societal understanding

Teachers face another barrier that goes beyond the classroom walls: a general lack of understanding in society about what neurodiversity is and what it looks like in a classroom. This isn't just an abstract problem; it directly impacts how they teach and the level of support they are able to provide to neurodivergent students.

Teachers are in the perfect position to make a difference in students' lives, but society often takes them for granted, assuming it's just their job to handle *everything*. That's not fair or realistic. Yes, educators need to be provided with information and resources. But every part of society that intersects with the classroom or a student has a responsibility to help shape young lives.

TIP

If you're a teacher, take a tip from the playbook of many neurodivergent students and exercise *self-advocacy* — the practice of speaking up for yourself and your needs. Understand what you need, then go after it. Sometimes it takes a teacher clearly communicating their needs for administrators to even realize that they need to act.

REMEMBER

Neuroinclusive teachers create the type of classroom where all types of minds can shine. They guide kids in how to understand themselves and their classmates, and they pull out creative approaches, so nobody feels lost or ignored. Teaching is not just about reading and writing; it's about making sure every student gets to show off what they can do. And that makes learning a whole lot more fun for everyone!

You only have so much energy and time

As a teacher, your time and energy is limited. That's natural, but just because a barrier is natural doesn't mean it isn't there. The Himalayan Mountains are a natural part of life, but you wouldn't be able to casually walk past them on a 40-minute lunch break one day. Now imagine that you *were* to try and do that — suddenly charge up one of Earth's tallest peaks on foot and on your own. You'd quickly hit the point of exhaustion and burn yourself out within just a few short miles.

You know, there's a reason why we have buses and trains, bikes and cars, scooters and spaceships. Much of where we need to go is too far and too challenging to try to get there on our own. That's why there are specialists who can help adapt curriculum to be more neuroinclusive, as well as resources out there for you to find. (For such resources, see the appendix.) Teachers *are* superheroes, but c'mon. Your powers don't include the ability to bend reality and time.

Coping and Succeeding as Educators

Whether through acquiring resources or self-care, you have to find the things that allow you to succeed. Following are a few strategies that can help get you there.

Understanding yourself and your needs

You know that safety briefing on the plane that you're *supposed* to pay attention to? Even if you don't, the one thing you probably consistently remember being told is to put your own oxygen mask on before assisting others. That applies to teachers as well. You *have* to understand yourself and your own needs in order to effectively teach. That way you're prepared to better assist your students when they seek help.

Teachers today carry many responsibilities far beyond just the subjects they teach. More often than not they shoulder these responsibilities without extra pay or time off. Balancing these responsibilities can sometimes lead to chronic stress, anxiety, and even burnout if teachers neglect their own needs. This can, in turn, affect a teacher's ability to manage time, stay organized, and maintain the energy needed to teach.

But fear not, the path to self-understanding and self-care is not a mysterious journey. It's about recognizing one's own needs and tending to them. Much like nurturing the seeds and sprouts of a garden, nurturing yourself and tending to your own needs allows you to be fully present and vibrant. That's the type of teacher that students need. By taking care of yourself, you can take better care of your students.

REMEMBER

This isn't about being selfish (can any teacher truly be selfish?). This is about being your best for your students. Educators who practice self-care activities such as managing a consistent daily routine that incorporates physical activity, meditation or quiet time, eating healthy meals, and sleeping regularly tend to excel in their duties. Hobbies and enjoyable activities, along with community support outside work, help teachers find joy, relaxation, and balance as well.

Understanding students and their needs

Just like you, your students have their own needs. All students learn differently, making it a challenge to support everyone. Utilizing practical tips for universal design for learning and inclusive education strategies helps educators understand students' skills, abilities, strengths, and needs.

TIP

As every student learns differently, supporting the unique learning styles of all students is a challenge. However, universal approaches and inclusive education strategies can assist educators as they support their students' unique abilities, skills, strengths, and needs. Following are some tips to help you broadly support diverse learners:

>> **Celebrate differences.** Understand that every student, whether neurotypical or neurodivergent, has unique strengths and viewpoints. Don't try to fit everyone into one mold; appreciate the richness of diverse brains. See neurodivergence as a valuable aspect of human variety, not a deviation. Encourage students to foster a safe space by showing kindness, dignity, and respect in all interactions.

>> **Vary teaching styles.** Enhance learning for diverse students by using various formats such as images, videos, and audio, and reflecting different cultural perspectives. Take Khushboo on our author team, for instance. She once watched a classroom session that followed days of students struggling to understand worksheets about human cell structure. In response, the teacher pulled out some art supplies along with a bag of candy. Students then constructed cell structures with the materials. Doing so suddenly allowed the lesson to click.

>> **Vary assessment methods.** Not all of us are good at standardized tests! Assessing students can include essays, projects, presentations, portfolios, artistic creations, peer assessments, discussions, quizzes, and informal check-ins. Recognize that some students may need extra time for assignments.

>> **Promote student interaction.** Group activities boost collaborative learning for diverse learners. Flexible grouping, driven by interests and strengths, eases socializing for neurodivergent students.

>> **Teach independence.** Giving options and chances to make decisions benefits students. Allowing kids to choose how they want to tackle an assignment, break it into smaller parts, and figure out what's most important to solve not only teaches them task management but also encourages them to be self-reliant. If someone's finding it tough, teachers can show them different approaches, gradually stepping back as confidence grows.

>> **Adjust the classroom environment.** Some students find busy classrooms with too many sensory stimuli distracting. Teachers can minimize sensory issues by observing student reactions to strong scents, bright colors, sound, and seating arrangements. Making adjustments to sensory sensitivities or creating a sensory-friendly corner in the classroom can make huge difference for neurodivergent students.

Teachers need to keep a close eye on their students to figure out their individual skills, strengths, and areas in which they need help. For instance, some students may do well in one activity but feel anxious about reading out loud. Others may get distracted easily. By recognizing these differences, a teacher can adjust how they teach to help students stay focused and comfortable. Whether it's giving a warning before calling on a student or offering extra support, understanding each student can lead to a more successful classroom for everyone.

Every student is unique, and their preferences and needs can require personalized responses. Even though there's no one magic method that works for everyone, certain tips and strategies can be helpful. These tools can help neurodivergent students excel in different grades and learning environments. Stay tuned, because in the next section, we dive into how specific settings can support neurodiversity, depending on a student's grade and developmental stage.

Empowering Students in a K-12 Setting

Your role in engaging neurodivergent students in the K-12 classroom is more than just a job — it's a life-changing mission. From the colorful days of elementary school, through the transitional time of middle school, to the challenging years of high school, your understanding, support, and connection can unlock potential like nothing else can. Teaching is not about just getting through the syllabus; it's about lighting a spark that lasts a lifetime. And believe us, that spark? It can change everything. So, grab your creativity and empathy; you've got minds to inspire.

Elementary school

Elementary school is where the magic starts, especially for neurodivergent students. As a teacher, you get to lay the foundation for their entire academic journey. These early years are filled with discovery and exploration, and your influence can spark curiosity and set them on a path to success.

TIP

By nurturing elementary students with self-awareness and self-acceptance, the neurodivergent child is better prepared to seek help, express their needs, and cooperate with other students.

Elementary school's a big new world for everyone, not just the students! You K-12 educators have the know-how to guide students on how to interact in the classroom and with their peers. Remember, every kid is climbing that steep learning curve at the start of school, and flexibility is key. Your adaptive techniques make all the difference in meeting each student's unique needs.

REMEMBER

These early years are prime for inclusive learning and multi-modal classroom fun. You hold the keys to meaningful education, aiding understanding of complex concepts. Universal approaches support all minds, whether or not a student's neurodivergent condition is evident.

Middle school

Ah, the thrilling roller-coaster journey of middle school — brace yourself for a wild ride! Few life changes are as jarring for our species as that transition from elementary to middle school. Who are these educators who *choose* to teach students flushed with complex physical, cognitive, emotional, and hormonal changes? In all, a pretty remarkable bunch. As a middle school teacher, your success in helping neurodivergent students thrive in your classroom begins before they even arrive.

As the elementary and middle school experiences are so vastly different, we've found that neurodivergent students often benefit from a transition activity during the summer before middle school that gently shakes up their routine.

Educators and school staff also have a role in this shake-up, and your part ensures that students shaken from their previous school experience can quickly and comfortably settle into your own classroom. Whether through open houses or campus tours, or your communication via email to parents and students, your impact on middle school students begins before the first bell rings.

That's sort of the secret here: Students have to learn *how* to be in middle school in addition to any subjects they learn. Remember, kids have a limited supply of focus and energy (okay, children have *a lot* of energy). If the learning curve in adjusting to middle school is too steep, that's less focus they have for the lessons you teach.

TIP

Ideally, you'd have the bandwidth to help all your students smoothly acclimate to middle school. But, just start with whichever homeroom (also known as a form class or tutor group) you're assigned. While some colleagues may see homeroom as just a place for attendance and announcements, this is your opportunity to shape students beyond what is taught in the curriculum. Remember that students talk, often sharing their opinions of their various homeroom teachers. If you create an impact on the students in your homeroom, that impact can spread.

Of course, acclimation and "learning how to be" in middle school — or "learning how to be" in general — isn't just limited to the first semester of the first year. Throughout their middle school experience, students are constantly figuring out how to interact and cooperate with others as well as learning how to function in our larger society. The more focus you place on making your lessons accessible for all types of learners, the more students will be prepared to venture out into the world ahead.

REMEMBER

Middle school is a time of self-discovery, where students navigate changes in themselves and in how they interact with others. Promoting inclusivity, staying vigilant against bullying, and fostering respect among students are a key component of helping your neurodivergent students thrive.

High school

It's an exciting responsibility to teach students as they transition from childhood into their adult life. High school teachers play a pivotal role in guiding students through increased independence, discovery of their interests, and preparing them for the challenges of work or higher education in the years ahead.

A key way educators can help neurodivergent students in high school is by helping them clearly identify campus resources and accommodations readily available to them. For students, a large part of high school is learning how to do more things on their own. Show your students what is available to them and how they can go about getting them. Then, provide a little guidance and encouragement to prompt them to seek these things out for themselves.

TIP

Many neurodivergent students also thrive with clarity and context. If given in a vacuum, standardized tests and future planning may seem like an unimportant chore to many neurodivergent students. However, by showing how these things concretely relate to the student and their own success, that little bit of context can help trigger enthusiasm within the neurodivergent mind.

High school teachers may have less time compared to middle school teachers to better understand each student's individual learning style. However, you can still support the learning styles of neurodivergent students by showing empathy and seeking to understand their perspectives and needs.

Fostering a positive school environment, creating a supportive space for different learning styles, and encouraging students to participate in activities that match their own interests can greatly benefit neurodivergent students.

Empowering Students in Post-Secondary Education

Post-secondary education can be challenging for neurodivergent students because most educational supports protected by an individualized education plan (IEP) are no longer in place (assuming the student was able to get an IEP at all). It's the student who must now be responsible for their own education, for navigating accommodations, and for building their own support networks and peer groups.

A lack of structure and clearly identifiable supports can lead neurodivergent students to forgo higher education or to drop out once enrolled.

Colleges and universities

College is a whole different world compared to high school. It is here where educators play a crucial role in the development of neurodivergent students. Educators are not just holding hands anymore; they're giving students a little nudge toward independence.

Let's be honest; depending on the size of your institution, you may not even know the *names* of most of your students, let alone any neurodivergent condition they may have. What can you do? Well, you can practice the accommodating techniques we discuss throughout this chapter. Not every strategy applies to the college and university level, but try to see what fits. In addition, you can be direct and clear with your students that you are open to approaches that work for them. Let them know that you're happy to discuss any accommodations and options they may need to succeed (see the example of Johnny Harris in Chapter 8).

Trade programs

Trade programs can be a great alternative to a university education for many. Areas such as nursing, machinery, graphic design, construction, cosmetology, and the culinary arts lend themselves to the type of hands-on learning acquisition on which many neurodivergent students thrive.

TIP

If you're a trade school instructor, keep an eye out for students who do things differently or in the "wrong" way. Divergent thinking patterns often lead neurodivergent thinkers to work out things differently than most. You'd be surprised how often this leads to the discovery of more efficient ways of doing things that you may want to adopt.

Boot camps and training courses

There will always be individuals who are eager to learn, but who aren't super pumped about enrolling in college or in a trade program. Because of the mismatch between traditional teaching and neurodivergence, it's safe to assume that neurodivergent individuals are well-represented in this group. But here's the thing: There are still opportunities to learn.

A trait among several neurodivergent conditions is the ability to acquire new skills through self-learning. We've seen many neurodivergent people learn new skills and trades simply by watching online videos and reading everything that they can. In fact, part of our own work is encouraging companies to look at the skills of applicants rather than just their formal education, as these self-taught learners often have stronger skills than most.

Boot camps — as popularized by the tech industry and now used in sectors ranging from the creative arts to design — are a great form of learning for many neurodivergent people. These hands-on training courses offer focused and intensive instruction in a short period of time. Perhaps you are a business leader who operates a boot camp or a training course as a way to identify new talent. You've probably never thought of yourself as an educator, but for many neurodivergent learners — you are!

Ongoing education

Neurodivergent learners are lifelong learners. Even after school's done, they keep on mastering new skills. Whether you offer driving instruction, piano lessons, human resources training, or leadership development, you've got neurodivergent students among those you teach. In fact, we're all learning new things from each other every day. Why not approach all of your interactions with other people in the most accessible and accommodating way?

Chapter **21**

Empowering Neurodivergent Employees and Colleagues

O ur workplaces need neurodiversity. Unfortunately, how our work spaces and policies are currently designed means that many neurodivergent employees are not fully supported. In this chapter, we discuss how neurodiversity has played an important role in human society, why workplaces need neurodiversity, and how to modernize workplace practices and policies to help neurodivergent employees thrive.

Appreciating the Role of Neurodiversity in the Workplace

Neurodiversity may be a new concept for some, or even for you (and that's okay! It's why we're here). However, neurodiversity is as old as humanity itself. Here's the interesting part: Even though neurodiversity has always been a part of society, for most of human history we were sort of clueless about it. That's now beginning to change.

We're now discovering the role neurodiversity plays in humanity, *and that's an exciting thing!* We're fortunate to be alive at a time when we're starting to understand it, learn its benefits, and realize how neurodiversity forms a crucial part of many communities, neighborhoods, and families. This new understanding includes recognizing the role that neurodiversity plays at work.

Brain variation and human progress

Before we go deep into exploring how neurodiversity fits in the modern workplace, first take a moment to appreciate how downright remarkable human brains are. These brilliant, intricate organs let us think abstractly, solve complex problems, communicate on a global scale, understand our speck of a planet as it relates to our vast universe, and *create art!*

Our brains have given us soufflés and supersonic jets. They've gifted us pyramids and poetry, comic books, carwashes, and curry. This lump of matter we each carry around in our head led us to develop agriculture, invent both the wheel *and* waffles, discover DNA, cure diseases, harness fire, walk on the moon, dive in the ocean, and enjoy *at least* 26 studio albums recorded by Cher. That's a lot! And there's so much more.

Here's where things get really cool — not all brains are built the same. It turns out that through all of human history our brains have come in variations called *neurotypes* (to understand neurotypes more in depth, check out Chapter 5). These neurotypes — such as attention-deficit/hyperactivity disorder (ADHD), dyslexia, and autism — have been a huge help in giving humanity some amazing things. Why? Well, imagine that we all had the same brains, all of us good at the same things and bad at the same things. We all may ace certain tasks, but we'd all face-plant when dealing with others. Doesn't sound like a great survival strategy now, does it?

Let's flip the script: Diverse brains means diverse skill sets, which means we're not all trying to hammer a screw because we can't figure out that there is a better tool for that task. Variations in our brains mean that people excel in different areas and observe different things, making our species more versatile. For example, many neurodivergent brains offer keen perception — ideal for hunter-gatherer survival. Others possess exceptional memory and focus, preserving vital knowledge. These variations in thinking drive human progress.

REMEMBER

Neurodiversity is just as essential to our modern workplace. We need a variety of skills and viewpoints if our organizations are to be innovative and productive.

From farming to firefighting to pharmaceuticals, our workplaces have relied on neurodiversity to innovate and discover new things — yet in our modern workplaces, we've done little to support neurodivergent employees. We can do better than that, can't we? Of course, we can. Instead of just waiting for the magic of neurodivergent thinkers to show up, why don't we get proactive? Let's switch gears and actively embrace neurodiversity at work! We can create a space where everyone's unique brainpower isn't just recognized, but the diverse ways our brains work are actively leveraged to make our workplaces stronger and more dynamic. Time to tap into the power of different thinking styles!

Neurodiversity fuels innovation

In modern corporate culture, the word *innovation* has been stretched, reformatted, and watered down to the point that it's now commonly used to mean all sorts of different things. But, what is it exactly?

To put it simply, *innovation* is the process of creating and implementing a new idea, method, or product that generates value. This can involve improving an existing product, developing a novel solution to a problem, or finding a new method of doing something. The key component of innovation is that it results in added value or improvement, whether that's increased efficiency, enhanced customer satisfaction, or a new market opportunity.

Among our author team, John spent several years assisting the United States government in implementing innovation practices that reduced spending, cut red tape, and sped government services. During this time, John's work was significantly inspired by Steven Johnson, a well-known science writer. One concept that particularly struck a chord with John was Johnson's idea of the "slow hunch." This idea suggests that groundbreaking ideas don't typically occur instantly. Instead, they develop slowly over time, starting as small hunches that gradually incubate and evolve into significant breakthroughs.

Why ideas need to collide

According to author Stephen Johnson, innovation most often results when the slow hunch of one individual collides with the slow hunch of another. Here's an example:

When Steve Jobs and his Apple team aimed to make an affordable and user-friendly personal computer, they lacked a crucial piece. A trip to Xerox gave them the missing link — an intuitive computer interface using icons and a mouse. Xerox had a great idea, but only used it in a few internal applications. Jobs and Apple combined this with their own ideas, resulting in the groundbreaking Macintosh computer.

REMEMBER

This example highlights how innovation often involves building upon existing concepts.

Fostering the collision

Author Steven Johnson and other top innovators promote the notion that the more our individual slow hunches have opportunities to collide with the slow hunches of others, the higher our chances of discovering truly innovative ideas. To facilitate this process, many engaged in innovation invest considerable time in designing workspaces and platforms that encourage people to come together and share their ideas, allowing for fruitful collisions of thoughts and perspectives. These collisions then serve as catalysts for groundbreaking innovations.

WHEN HUNCHES COLLIDE

It was this idea of slow hunches colliding that inspired John and his colleague Matt Collier to take a dusty storage room located in the basement of the U.S. Office of Personnel Management and transform it into the nation's first federal innovation lab.

"Even that idea of creating a place to collide ideas was itself the result of colliding ideas," says John. "I still remember the day that Matt walked into my office, closed the door, and said, 'Hey, I have an idea.' It turned out that I had one too. That got us talking, and it eventually led to the development of our agency's innovation lab. The funny thing is that neither of our ideas had anything to do with the creation of a lab. Yet, by colliding our 'hunches' together, it eventually led to just that."

The innovation lab that the agency built, now The Lab at OPM, inspired the creation of innovation labs across the federal government. Each of these labs is now used to collide ideas and create solutions that save money and improve public services. One of these labs is the one located within the U.S. Government Accountability Office (GAO), the independent agency tasked with auditing and investigating federal spending and programs. The GAO predicts that the ideas emerging from its own lab will "save billions of dollars for taxpayers" in the coming years.

How neurodiversity intersects

Fostering innovation is not just about the types of physical spaces we create; the diversity of perspectives within a team plays a vital role in speeding up the collision of ideas. When individuals from different backgrounds and experiences collaborate, they bring together a much larger pool of potential ideas compared to a homogenous group. *That's why diversity in our workspaces is a key driver of innovation.* The collision of diverse hunches increases the likelihood of sparking new and innovative concepts, all thanks to the richness of our varied perspectives.

If you take that reality one step further — incorporating neurodiversity into the workplace is a powerful extension of this — neurodivergent individuals, with their unique ways of thinking, perceiving the world, and processing information, contribute significantly to the richness of the diversity of perspectives within teams. When you create an inclusive environment that embraces neurodiversity, you open up even more opportunities for diverse ideas to collide and mash together. Increased innovation is the result.

Neurodivergent team members may approach challenges and tasks from unconventional angles, leading to solutions that may not have emerged in a neurotypical-dominated setting. Neurodivergent brains work differently, bringing fresh insights and approaches to problem-solving. This opens up a whole new realm of innovative possibilities. By creating an environment inclusive of neurodiversity and valuing the contributions of neurodivergent team members, you unlock untapped creativity and talent and propel our workplaces to new levels of success.

You may be wondering whether diversity-driven innovation is applicable only in office-based workplaces. Walgreens has created structured environments in distribution centers that suit the skills of neurodivergent individuals, leading to increased efficiency and accuracy. The supermarket chain Asda has introduced "quiet hour" initiatives, influenced by insights from neurodivergent employees, to create a more comfortable shopping environment for all customers with sensory sensitivities.

Embracing neurodiversity in the workplace doesn't just enrich our culture — it propels innovation. It's fascinating how the power of diverse minds shapes our world, right?

Updating How We Work

Now that you have seen how neurodiversity accelerates innovation, in the following sections you look at ways to foster neurodiversity inside the workplace. But first, we take a few moments to outline the landscape of where most workplaces are now.

Understanding how we got where we are

Right now, many workplaces are designed around what works best for people who think in typical ways. Take the usual 9-to-5 schedule, for example. It expects employees to be switched on and productive during set hours. But not everyone's brain is on the same timetable. Early birds, night owls, and folks with conditions such as ADHD may find this routine tough to stick to. Yet, most workplaces don't cater to these different rhythms.

Then, think about physical office spaces. They're often planned for efficient use of space or to boost certain types of chit-chat. The problem? Not everyone can work effectively under these conditions. Autistic people, for instance, may find the usual buzz of the office overwhelming.

Lastly, consider how people interact at work. Often there's a focus on constant teamwork, networking, and group decisions. It works for some, sure, but not for all. Those with neurodivergent thinking and communication, or simply those who are introverted, may prefer more solo work or talking via email instead of face to face.

So, modern workplaces — with their set schedules, specific office designs, and certain ways of communicating — are pretty much tailored for certain neurotypical folks, but not for all. And that's where we need to do better.

Over time, workplaces have developed certain attitudes, rules, and customs that are deeply rooted. Often, employees don't even think about why these attitudes and rules are there.

Here's a bright spot: More and more, people are recognizing the importance of neurodiversity in the workplace, and businesses are beginning to understand that they need to be more adaptable and welcoming to everyone. We're starting to see more remote and flexible work options, which can be a game-changer for some people with different neurological makeups. In addition, there's a growing awareness of mental health and the fact that employees are not all wired the same way, which is great. Sure, there's plenty of road ahead, but we're making progress toward creating workplaces that value and include all kinds of thinkers. That's a really exciting shift, don't you think?

Adjusting workplace mindsets

One of the most impactful ways you can make a workplace more neuroinclusive is by adjusting your mindset around neurodiversity. Neurodiversity should be an important part of work culture. Here are some tips to help move toward that place:

» **Celebrate diversity.** Recognize that each employee, neurotypical or neurodivergent, brings their own unique strengths and viewpoints. Forget about trying to fit everyone into the same box. Instead, appreciate the variety and richness that comes from different types of brains.

» **Educate, educate, educate.** Make sure employees and managers understand what neurodivergence means and what neurodiversity means for the workplace. This education can involve training programs, workshops, or sharing resources and stories. The more we all understand, the better we can support one another.

» **Rethink workplace processes.** Are there things you do just because "that's how we've always done it"? Maybe it's time for a rethink. We share some fresh ideas in "Updating workplaces processes" later in this chapter.

» **Offer support when needed.** Support may involve giving people access to assistive technology and mental health resources or just making sure that managers are understanding and supportive of neurodivergent employees. Additional examples include the use of noise-canceling headphones in noisy work environments and a chair to sit on for a cashier in retail stores.

» **Don't forget families.** Employees who are parents, caregivers, or family members of neurodivergent children often have needs not met by workplace policies. These needs may include flexibility in scheduling, modern time-off policies, understanding from management, a supportive company culture, and comprehensive health insurance benefits. Even employees who may care for neurotypical children often need these things too! (Turn to Chapter 23 to learn what other ways you can support the parents and caregivers of neurodivergent children.)

» **Connect employees.** Linking neurodivergent employees can be valuable. It fosters socializing, idea sharing, and professional growth, and develops strengths. Initiatives such as a neurodiversity employee group or an awareness week can make a big difference.

TIP

It can be incredibly helpful to connect neurodivergent employees not only with each other but also with colleagues who parent or care for neurodivergent children. Whether they're neurotypical parents or neurodivergent themselves, both groups of employees suddenly find mutual understanding and support from each other.

REMEMBER

Keep in mind, we're not talking about extra perks or special treatment here. Embrace diverse minds, educate about differences, change routines, be flexible, and support one another. Those are things that help everyone thrive.

Updating workplace policies

Of course, it's not just employee mindsets that need to be adjusted. Workplace policies need to be shaped into something more modern as well. From local landscaping companies to international airlines, every workplace has its own set of operating rules. Some of these rules are official policies, those things written down that state what is expected of employees, what is acceptable, what is prohibited, and how an organization likes to get things done. Others are unwritten rules, what some call a "hidden curriculum." Such unwritten rules include attitudes toward work-life balance, power dynamics, norms around how breaks and time-off should be taken, and "how the boss likes things to be done."

Following are tips on how to modernize official workplace policies so that they are more inclusive of neurodivergent colleagues:

>> **Flexible work arrangements:** Offering work hours that can be changed or remote work options is a good idea. Some people may need a late start because of their medication timing, or they may perform better with frequent short breaks. Adopting rules that support these flexible work styles is a win, as everyone does their best work differently.

>> **Adjusted workspaces:** A busy, open office isn't for everyone, right? An individual with ADHD, for instance, may benefit from a standing desk or a quieter area. Others may need particular lighting or noise conditions. It's wise to have policies allowing these modifications, helping employees be more comfortable and productive.

>> **Updated training resources:** This may mean offering training materials in different formats (such as videos, texts, or audio) to suit varied learning styles, or providing extra time for training activities. Extra training for managers and staff on understanding and aiding neurodivergent employees is also a great move.

>> **No tolerance for discrimination:** Nondiscrimination policies should specifically mention neurodivergence, ensuring that neurodivergent employees get the same rights and protection as everyone else. If any employee feels they're being unfairly treated, they should have a simple, secure way to report it and seek help.

It's essential that these policies aren't just lip service — they need to be implemented and enforced to make a real difference.

Modernizing workplace processes

A workplace process is simply a well-structured set of steps that guide you through getting tasks done in a smooth and efficient manner. This can include the employee onboarding process, conducting a performance review, how things are purchased or counted, how teams are managed, and how to engage customers. Here are some specific ways you can update workplace processes to better support neurodivergent colleagues:

>> **Clear communication:** Make communication simple and jargon-free. Use different formats to share information. Post workplace norms and rules in breakout rooms. Consider sending meeting agendas beforehand or using visuals in presentations. These things can be a big help for those who communicate and process information differently.

>> **Single-tasking:** Many workplaces love multitasking, but it's not for everyone. Encourage focusing on one task at a time or let employees manage their tasks their way. This can be great for someone with ADHD who works best when diving deep into one particular task at a time.

>> **Collaboration that works for all:** Group projects and brainstorming are awesome, but remember those who work best solo or need more time to process and ideate. After brainstorming, send an email for extra thoughts. It'll help those who need extra time.

>> **Fair feedback:** Offer clear, helpful feedback that respects individual communication styles. Evaluate performance considering the various ways employees shine, not just traditional success markers.

>> **Regular check-ins:** Hold regular forums for employees to voice their needs, worries, as well as solutions. It helps the workplace adapt and cater to everyone's abilities.

Revamping workplace processes for neurodivergent inclusivity doesn't mean a complete overhaul — it's about flexibility, understanding, and prioritizing people. The goal is to create a workspace where everyone can work their best way.

Employing Affirming Practices

In the next sections we dive into practical strategies for supporting and empowering neurodivergent colleagues.

Exploring the complexities of cognition

Cognition is the term for how we think. At this point in the book, it likely isn't a shock to learn that neurodivergent brains think differently (for a deep dive on these differences, make sure to read Chapter 6 through 9). As workplaces have been traditionally designed for neurotypical thinkers, there is ample opportunity to reshape workplace practices so that they support all variations in employees' cognition.

REMEMBER

Many neurodiverse individuals, particularly autistic colleagues, are known to interpret language in a very literal way. They tend to process information exactly as it's stated, without reading into implied meanings.

Take Alan, for instance. You may be wondering why he always skips group lunches. Well, have you ever specifically invited him to lunch, or was an invitation just implied through the body language and behavior of you and your coworkers? If so, Alan may be unsure as to whether he should join you or not.

Now, what if you made the effort to communicate more clearly? We bet that would help a lot. Imagine going up to Alan and saying, "Hey Alan, we're going to grab lunch. Would you like to join us?" Being clear about your intent would clarify things for Alan and make him feel comfortable joining the group.

THE CONCEPT OF "GOOD ENOUGH"

Neurodivergent people often use perfectionism as a coping mechanism. They may use it to aid their cognition, or to deal with criticism that comes from being told their way of thinking and doing things isn't "the right way." Some neurodivergent people, particularly those with autism or ADHD, also just *enjoy* the process of refining something again and again to make it better and better.

Striving for excellence is valuable, but unchecked perfectionism can cause missed deadlines and hinder progress. Balance is crucial, and using mechanisms for task balance is essential for success. The concept of *good enough* is one such mechanism that we've found.

We discovered this mechanism while collaborating on a series of presentation slides. John really enjoyed continuously refining the slides to ensure that the word choice, graphics, and colors would connect well with the audience we were to present to. But, we also were facing a strict deadline that conflicted with the empathy for our audience that John was pouring into the project. After several revisions, Ranga happily stated that the slides were good enough.

With his autistic and ADHD brain, good enough wasn't a concept that was familiar to John. A bit confused, John asked Ranga what that meant. Ranga replied that for neuro-typical thinkers, many projects or products don't have to be perfect. They just need to be good enough to use or to convey an idea.

"I remember immediately going out and asking neurodivergent friends if they were familiar with the concept of 'good enough'," says John. "Like me, it was totally unfamiliar to many of them. But, you know what? A lot of us began to use it as an accommodation in our work. Checking our work at regular intervals to ask, 'Is this good enough?' shaved hours off our work and allowed us to move on to other tasks. It's not about producing something that is substandard. It's about realizing that you've reached an acceptable point for others and saying 'Okay, I can move on'."

Supporting communication differences

Neurodivergent people often have communication styles that differ from neuro-typical people (we go into these differences in Chapter 18). Here's a few tips to help your cross-communication between these styles go much more smoothly:

>> Be patient and attentive, allowing colleagues to express themselves fully.

>> Offer written communication options such as email or messaging to provide additional clarity and time to process information. Ask the person their preferred method of communication.

>> Provide a quiet and private space for discussions to minimize sensory distractions.

>> Offer flexibility in meeting formats and consider using visual aids or diagrams to enhance understanding.

>> Recognize, allow, and accept the slower pace of exchanges when an augment-ative and alternative communication (AAC) device is being used.

TIP

Imagine a gathering or meeting where everyone is participating except for one person, Amara. She sits quietly. If you're leading the meeting, you can directly address her, saying, "Amara, we'd love to hear your thoughts on this matter." This way, you give her the chance to join the conversation without her feeling the pressure of trying to figure out the right point to jump into the discussion.

TURN-TAKING: WHEN TO JUMP IN

There's a stereotype out there that neurodivergent people — particularly those with autism or ADHD — may interrupt or interject themselves into conversations at points where others don't find it appropriate for them to speak. Due to communication differences, this does happen at times.

However, we'd argue that neurodivergent people are just as likely (if not more likely) to *not* join a conversation due to fears of misreading social cues. This can present challenges in workplace meetings, as the fear of being judged can make it difficult to determine when to speak up.

Turn-taking in group conversations or meetings refers to the practice of participants speaking one at a time in an orderly and respectful manner, ensuring everyone has a chance to contribute. This can produce anxiety in many neurodivergent colleagues. John from our author team describes this anxiety as being similar to the hesitation you may have felt in your school days deciding when to jump in to a game of Double Dutch jump rope — chose the wrong moment and you may stumble and trip!

It's worth noting that turn-taking challenges aren't exclusive to neurodivergent individuals — societal contexts can create parallel situations. For example, being the only woman, person of color, or immigrant in a group can generate pressures that cause self-doubt and hesitance to voice one's thoughts.

Here is a personal example from Ranga, whose son often seems disinterested during family dinners, doesn't join in the banter, and appears aloof. But Ranga has realized it's just his son's way of processing the conversation and what he hears. Whenever Ranga pauses to ask his son for his thoughts, he regularly replies with deep insights.

In addition, keep in mind the following two tips to help support communication differences between you and your colleagues:

> **Be aware of unwritten rules (what we earlier called the "hidden curriculum" of work).** Make an effort to communicate expectations clearly to all employees.

> **Be thoughtful to accommodate for auditory delays.** Some neurodivergent individuals may experience auditory delays, which means that it takes a bit of time for what they are thinking to come out of their mouth. That's okay. We can adjust for that. Keep this in mind in situations such as job interviews, group meetings, and one-on-one conversations.

Supporting executive functioning

Executive functioning is how our brains remember information, plan our day, set goals, and navigate obstacles and distractions. It involves managing emotions and behaviors, employing strategic planning, and time management. Neurodivergent people use executive functioning different than most. In workplaces that are mostly designed to support *neurotypical* executive functioning, this can be a challenge.

Here are some strategies we find helpful in supporting executive functioning for all employees:

» **Clear instructions:** Provide clear instructions for tasks and projects, breaking them down into smaller steps if needed.

» **Visual aids:** Use charts, checklists, or timelines to help organize information and set priorities.

» **Reminder systems:** Use calendars, alarms, or task management apps to assist with deadlines.

» **Work breaks:** Provide breaks to help manage attention and avoid being overwhelmed.

» **Goal setting:** Set realistic and achievable goals and track progress regularly.

» **Positive feedback:** Reinforce good executive functioning practices.

» **Supportive environments:** Create an environment that enables employees to feel comfortable seeking help.

TIP

Context can also help tremendously with executive functioning. At a major university, we ran several neurodiversity awareness sessions for the staff. Later, we received feedback that managers had followed our advice and had started to give their employees more detailed context about why they were being assigned certain tasks. As a result, tasks that were previously seen as busy work were now passionately embraced. It had a massive impact on employee performance and project outcomes.

Leveraging neurodivergent strengths

Throughout this book we discuss common strengths associated with neurodivergence (to learn about this in more detail, check out Chapters 6 through 10). It's important for workplaces to utilize and leverage neurodivergent strengths — otherwise they're missing out on a lot.

Unlocking the power of people who think differently comes down to understanding their unique strengths and adapting to them. Here's how you can do that:

>> **Allow flexibility in workflow:** Let employees manage their work in ways that work well for them, and you may see a boost in productivity and job happiness.

>> **Change things up:** Sometimes people who think differently thrive on variety and find same-old, same-old tasks a total snooze. Swapping tasks around can keep them tuned in.

>> **Find their sweet spot:** Neurodivergent people can be superstars in certain areas, such as in spotting patterns, getting down to the details, or comping up with out-of-the-box ideas. Find roles that tap into these talents for a win-win situation.

>> **Have a guide on the side:** Offer a mentor who can help decode office rules, answer questions, and offer advice. This should be someone who gets the individuals' unique needs and can serve as a translator between them and the team when necessary.

Call mentors of neurodivergent colleagues *mentors*. Don't use condescending terms like *buddy* that some businesses have used in the past.

Neurodivergent employees can be fantastic mentors for new hires, whether they're neurodivergent or not.

For example, John on our author team saw this in action while visiting the headquarters of National Australian Bank (NAB) in Melbourne. There, staff who were either neurodivergent or parents of neurodivergent children, willingly took on the mentor role, offering their wisdom and experience to help autistic interns and new hires. We think that's a great model other companies should adopt!

Adopting inclusive recruiting practices

Recruiting is the crucial process of seeking out and attracting a wide range of qualified candidates who align with your organization's values and goals. By deliberately adopting inclusive practices from the outset, you set the foundation for a diverse and talented workforce that can drive your company to new heights.

Here's a list of key practices to attract neurodivergent talent in a straightforward way:

>> Focus on building an organization where everyone has equal opportunities to grow. Include people with disabilities not only as team members, but also as leaders.

>> Show your inclusive culture in your branding. Teach everyone in your company about neurodiversity to create a supportive atmosphere.

>> Consider creating employee resource groups (ERGs) for neurodiversity. They offer support and encourage inclusive conversations and professional growth.

>> Develop a clear list of your job types and their requirements. Each job type should plainly lay out the required qualifications, skills, and experiences needed. Go one step further and talk about the strengths one needs to really excel in each job type.

>> Hold regular information sessions for potential candidates. Show them what you are all about — your mission, your culture, the types of positions you offer, and of course, your intentional disability inclusion and neuro-diversity efforts.

>> Be up-front about the accommodations you proactively offer. Make sure accommodations are a part of every step of the recruiting process. Websites such as askjan.org and inclusively.com can help.

Implementing inclusive hiring practices

After the initial recruiting phase, it's time to select the best candidate from the pool. This involves conducting interviews, skills assessments, reference checks, and salary negotiations before making a job offer. Now, when it comes to hiring neurodivergent individuals, consider these ideas to make the process less intimidating and more inclusive:

>> **Job descriptions:** Keep them simple, divided into "required" and "desired" skills. Highlight any available supports.

>> **Feedback:** Provide resume feedback through hiring managers or applicant tracking systems.

>> **Interviews:** Focus on job-related skills over social competence or culture fit. Be open to second chances — nobody's perfect!

>> **Interview preparation:** Share clear details about location, length of time, and who is conducting the interview. Go over questions, the evaluation process, and next steps.

>> **Interview environment:** Use a quiet, low-lit room, avoiding crowded spaces or distractions.

>> **Communication:** Respect communication differences, such as varied eye contact or answer speed.

>> **Questions:** Ask concrete questions, avoid abstract concepts, and focus on specific experiences. Ask follow-up questions as needed.

>> **Offer alternatives:** Not everyone thrives in a traditional interview setup. but they may have other ways of demonstrating their experience and worth.

A mindful and inclusive approach to hiring can return big results.

Demystifying accommodations

Believe it or not, we all use accommodations. An accommodation is simply something that helps us do things smoothly and efficiently. At work, accommodations can be as simple as having an ergonomic chair for better posture or using time management techniques to stay organized. For restaurant cooks, it may mean using anti-fatigue mats to ease the strain of standing. And yes, even having the flexibility to handle family emergencies, such as taking care of a sick kid, is an accommodation.

We all require accommodations to thrive in our jobs. When it comes to work, making accommodations for neurodivergent people can help them perform at their best.

Here's a list of approaches we find helpful when it comes to offering workplace accommodations:

>> Keep an updated list of accommodations you readily provide and proactively offer workplace adjustments to all employees, not just upon disability disclosure.

>> Make sure your disability disclosure process, as well as your accommodation request process, is simple to use and easy to understand.

>> Offer flexible work schedules, which can often increase a neurodivergent person's productivity.

>> Provide stim tools, which can help both the neurodivergent employee as well as everyone else.

>> Give meeting notices in advance and ask about communication preferences.

>> Record meetings when possible for later study and note-taking (check legal requirements).

>> Give employee co-ownership of projects, tasks, and problems to solve. Doing so combines complementary strengths of co-owners and engages the attention of the employee, which fits many neurodivergent minds quite well. This book is an example of co-ownership in action!

REMEMBER

All humans use accommodations, not just neurodivergent and otherwise disabled employees. By proactively providing workplace adjustments that allow people to work better, you don't just have to rely on a disability disclosure process to get employees the accommodations they need. It also deflates the fear around workplace accommodations and instead makes it a positive part of workplace culture.

Leveraging resources for employers

No need to worry if you're feeling lost while trying to support neurodivergent colleagues. You're not alone. Check out the appendix for tips. Or for a quick solution, ask your neurodivergent coworker about the things that help them work best. Most likely, they'll tell you!

6
Building a Neuroinclusive World

Discover the concept of universal design — merging equitable use and empathy into design to create things that work for everyone — and how applying universal design principles into products fosters neuroinclusion.

Find out how to support parents and caregivers of neurodivergent children, including incorporating neurodivergent families into daily life.

Look at what it means to take a neurodiversity-informed approach to societal development and how to build a neuroinclusive world that works for everyone.

Chapter **22**

Designing for Universal Inclusion

When buying scissors, most people probably grab a pair without a second thought. But for the 10 percent of us who are left-handed, it's a different story. Traditional scissors cater to right-handers, making them difficult for lefties to use. While there are scissors designed specifically for left-handed people, the catch is that they're harder for righties to use. A solution? Scissors that work for *both* hands. Such inclusivity in design, creating things that work for everyone, is called *universal design*.

Imagine a world where everything, from architecture to technology, is designed for all. Envision a hassle-free visit to renew your driver's license where you are greeted with a pleasant environment, a straightforward and easy-to-understand process, with minimal wait times and little paperwork to fill out. This is the essence of universal design: making stuff friendly, easy to use, and accessible for all of us. In this and the following two chapters, we delve into how we can create a neuroinclusive world, beginning with how universal design can make every brain type feel at home.

Adopting Principles of Universal Design

For a truly inclusive world for neurodivergent individuals, universal design is essential. Without it, we risk continually wasting both time and resources trying to fix flawed systems. Universal design ensures inclusivity from the start and helps correct our past mistakes. It guarantees accessible spaces and experiences for everyone. Beyond just addressing needs, it enhances community participation and fosters belonging for all.

WARNING

It's important to remember that universal design does not address every need for every person in every possible situation (an impossible thing to do). What it focuses on is addressing the most needs for most people in as many situations as possible.

Scoping the principles

Seven key principles are at the heart of universal design.

>> **Equitable use:** The same design should be useful for everyone.

>> **Flexibility in use:** Designs should cater to diverse preferences and abilities.

>> **Simple and intuitive use:** Designs should be easy to understand, regardless of experience or ability.

>> **Perceptible information:** Information should be clear and noticeable.

>> **Tolerance for error:** Designs should minimize hazards resulting from accidental misuse.

>> **Low physical effort:** Designs should be able to be used comfortably with minimal fatigue.

>> **Size and space for approach and use:** Everyone should fit in and feel good.

Infusing empathy in design

You know that feeling when someone just gets you? That's empathy. Now blend that with design and you've got universal design. It's like a product or service giving you a reassuring nod, saying, "I understand what you need. Here you go!" By truly tapping into users' actual needs, especially those often overlooked, designs become more inclusive and welcoming. But without that touch of empathy, without stopping to ask "Wait, does this work for everyone?" we risk creating things that unintentionally exclude. And no one wants that, right?

REAL-WORLD EXAMPLES
OF UNIVERSAL DESIGN

The concept of universal design wasn't just plucked from thin air; it was crafted by some really smart folks, including architect Robert Mace and his colleagues at North Carolina State University. Another key player was author Selwyn Goldsmith, who introduced the concept of the *curb cut effect* — when things designed for disabled people end up benefiting lots more. Here are some examples of universal design — including the curb cut:

- **Ramps and sloped entries:** Originally designed for wheelchair users, ramps and sloped entries also benefit parents with strollers, delivery people, those with crutches, and even skateboarders.

- **Closed captioning:** While initially created for the deaf and hard-of-hearing community, closed captioning is widely used in noisy environments or by those learning a new language.

- **Lever door handles:** Instead of round knobs, lever-style door handles are easier for individuals with arthritis, those carrying items, or individuals with limited grip strength.

- **Voice-activated assistants:** Devices such as Amazon's Alexa and Google Assistant help the visually impaired with internet access and task management, while also benefiting users such as cooks with messy hands or drivers on the road.

- **Tablets:** Initially launched as a device for web browsing, photo viewing, ebook reading, gaming, and email, tablets are now used by doctors for medical records, pilots for navigational charts, and nonspeaking autistic individuals for communication.

- **Adjustable font sizes on websites and apps:** This feature assists low-vision users, but it's also handy for users on various devices or those who prefer larger text for reading.

- **Curb cuts in sidewalks:** Initially designed for wheelchair users, curb cuts also assist bicyclists, skateboarders, and pedestrians with carts or suitcases.

- **Wide doorways and hallways:** While designed for wheelchair accessibility, wider doorways and hallways ease movement for people carrying large items or moving in groups.

- **Audible crossing signals:** Essential for low-vision individuals, these signals also help children, the elderly, and anyone distracted or unfamiliar with an area.

- **Ergonomic product designs:** Products designed with ergonomic considerations not only cater to individuals with physical disabilities, but also promote comfort and efficiency for all users.

- **Multi-level kitchen counters:** These counters cater to individuals of various heights, including children, adults, people of small stature, and wheelchair users.

Consider the design of a public park. A standard design may have benches, play areas, and walking paths. Infusing empathy into this design would mean thinking of parents with babies and adding changing tables to restrooms. For those with mobility challenges, paths should be even with frequent shaded seating spots. For blind and low-vision guests, tactile paving on pathways and braille descriptions near landmarks can be beneficial. By addressing varied visitor needs, the design naturally becomes more inclusive and inviting.

User empathy is at the heart of universal design — a heart that cares about everyone.

Benefits of universal design

Universal design creates accessible and inclusive environments, benefiting everyone by ensuring ease of use, eliminating barriers, and accommodating a wide range of individual needs and preferences. Unexpected benefits of universal design include increased market appeal due to broader usability, reduced need for future modifications (saving costs and resources), and fostering a sense of community.

A real-life example of universal design that we love is The Kelsey, a housing development company cofounded by both disabled and nondisabled individuals, that emphasizes inclusive and affordable housing. They create apartment buildings suitable for both disabled and nondisabled residents. Besides being physically accessible and sensory-friendly, their staff don't just handle packages or suggest restaurants — they're trained to assist residents with physical and cognitive needs.

Putting Universal Design into Practice

How do you apply the universal design principles to neuroinclusion? You've got to start with listening to the users to develop that deep user empathy — and we mean all types of users! We walk through how to do that in the following sections.

Centering and amplifying voices

In universal design, it's not enough to include divergent voices and opinions. A key component is making sure that everyone's voice is centered and amplified. Imagine you're at a social gathering and everyone there has been invited to the party. That's inclusion. But, in the middle of the party there's a compelling storyteller. They're telling the most amazing stories, but their voice is being drowned out by the louder crowd. This storyteller represents the neurodivergent voices in our society — they have so much to offer, but are often drowned out.

When we say that every voice should be *centered*, that means giving that storyteller a stage. It's similar to ensuring neurodivergent voices aren't just in the room but are also helping steer the conversation. And *amplified*? That's like handing someone a microphone and a spotlight to make sure that what they are saying can truly be heard and seen by all.

Ensuring neuroinclusive design

Beyond universal design, we advocate for neuroinclusive design, ensuring that our environments, products, and services meet diverse neurological needs. As a result, companies increasingly see neuroinclusive design as business-smart, with Apple leading the way. The company not only incorporates the input of neurodivergent users in product design but also collaborates with neurodivergent individuals on best practices, which it shares with external app developers for its store. In addition,

» Microsoft's Learning Tools for OneNote offers features such as the Immersive Reader, which aids reading and comprehension. Users with dyslexia find it especially beneficial, as it adjusts text spacing, provides visual and auditory feedback, and more.

» The Google Read&Write tool offers text-to-speech, dictionary, and picture dictionary features. This allows ideas to be communicated in multiple ways.

» The Brain in Hand app uses simple, intuitive icons and reminders to help users manage anxiety, remember tasks, and cope with difficult situations. This can be helpful to autistic individuals.

» Many ebooks and e-learning platforms offer multiple modes of engagement. Users can read text, listen to audio versions, or even change background colors to reduce visual stress — features that particularly support individuals with ADHD and dyslexia.

Universal design isn't limited to just technology. In Melbourne, Australia's Neurodiversity Hub introduced our author team to sports facilities, from football stadiums to cricket grounds, that offer sensory-friendly seating. La Trobe University in the Melbourne suburb of Bundoora amplifies the voices of its neurodivergent community through its Neurodiversity Project. This initiative enhances campus accessibility, and students can intern, earning credits for their degrees.

WARNING

When designing for neurodivergent individuals, always center and amplify neurodivergent voices. Many products meant for neurodivergent individuals fail not due to lack of intent or investment, but because creators didn't seek their perspectives first. Prioritizing the end-user seems straightforward, yet it's frequently overlooked.

TIP

The next time you hear about a design or initiative for neurodivergent people, ask "Whose voices are guiding this?" Because, trust us, it makes a big difference.

Here's an example of the benefits of neuroinclusive design: Researchers designing a classroom app for assignment tracking sought feedback from both neurodivergent and neurotypical students. An autistic student recommended a notification system where users can choose alert types, from soft vibrations to clear visual cues. Incorporating this suggestion made the app accessible to those with sensory sensitivities and also won acclaim from neurotypical users for its adaptability and ease of use.

Practicing neuroinclusive design

Of course, you can't change the world by yourself, but here are some practical ways you can help create a more neuroinclusive world:

>> **Use plain language in written and verbal communication.** Visual aids, symbols, and straightforward layouts can help many understand information better.

>> **Offer flexible working or learning environments.** For instance, allowing for varied lighting or seating options can accommodate different sensory preferences.

>> **Collaborate with neurodivergent individuals.** If planning something, involve neurodivergent individuals in the process.

>> **Regularly seek feedback from neurodivergent individuals.** Ask what changes or additions would make things more inclusive for them.

>> **Take sensory considerations into account.** In communal spaces in offices or homes, consider having quiet zones or sensory rooms where individuals can retreat if they're feeling overwhelmed.

>> **Plan events with inclusivity in mind.** Provide clear schedules and quiet spaces, and avoid sudden loud noises or flashing lights that may be distressing.

>> **Make use of clear, consistent, and easily identifiable signs in public spaces.** This assists individuals, especially those with processing differences, in navigating spaces confidently.

>> **Include all voices in design considerations.** Insist that anything designed for neurodivergent users is designed with neurodivergent users.

Chapter **23**

Supporting Parents and Caregivers

D id you know that there are five fundamental requirements for human survival? You can likely list a few immediately: air, water, and food to sustain us. Equally significant are the remaining two: shelter and social connection. Shelter not only shields us from the elements but also offers security, privacy, and a sense of safety. As for social connection, humans are inherently social beings, and it is essential for our physical, mental, and emotional well-being that we build and nurture relationships with others.

Regrettably, parents and caregivers of neurodivergent children often experience isolation from those around them. This lack of social connection arises because many people aren't familiar with neurodiversity and the unique aspects of a child's neurodivergent condition. When confronted with this unfamiliar territory, it's a natural human response to feel uncertain about how to respond. This uncertainty can sometimes cause well-intentioned individuals to withdraw, hesitate in offering support, or overlook the inclusion of families with neurodivergent children in various aspects of their daily life. In this chapter, we delve into one of those five fundamental needs — social connection. We focus on the obstacles that parents and caregivers of neurodivergent children encounter in fulfilling this need and how you can help.

Practicing Empathy

Parenting a neurodivergent child is a huge responsibility and a beautiful and complex undertaking. Many caregivers report feeling completely lost, overwhelmed, and isolated in their parenting journey. Those who care for neurodivergent children not only deal with the usual challenges of parenting, but they also have the added responsibility of navigating a world that often doesn't cater to their children's unique needs. This requires these caregivers to scrounge about for information on their child's condition, seek out and maintain adequate services and support, confront social stigma and judgment, advocate for the needs of their child and their family, and juggle everything else in their personal and professional lives. *That's a lot!*

If you're a person who knows one of these parents, you may be unsure how to offer meaningful support. Perhaps you're hesitant to ask too many questions or fear crossing boundaries. Maybe you feel that you don't know enough or are afraid of saying the wrong thing. You may even feel some guilt about wanting to assist but not knowing where to start. If so, don't fret. Such feelings are entirely normal.

So, how do you get around all this so that you can provide meaningful support? Well, the best place to start is with *empathy.*

Empathy is a remarkable asset when it comes to supporting parents and caregivers of neurodivergent children. It's like a bridge that connects you to their world, enabling a deep understanding of their unique experiences, joys, and challenges. Understanding someone from their perspective helps clear away your own uncertainties, fear, and anxieties. As a result, you become better positioned to offer support.

REMEMBER

To practice empathy, start by listening without judgment, taking time to understand the other person's perspective without inserting your own. Step into their shoes and try to genuinely sense what they feel.

Understanding the parent as a person

It's easy to overlook the fact that our parents are human beings just like us. They face their own challenges, possess unique mindsets, behaviors, habits, and worldview. The monumental task of raising a child demands extensive preparation, ample support, and an abundance of patience.

If you're a parent, you know that your identity encompasses much more than just being a parent (though that's undeniably a significant part of it). Similarly, when you have friends who are parents, you see them as multifaceted individuals as well. Yet, a quirky thing happens to those who parent neurodivergent children.

Often, those around them stop recognizing their complexity and see them solely as parents of neurodivergent children. It's as if everyone has suddenly forgotten who they are beyond their parenting role.

TIP

One of the best things you can do to support someone who is raising a neurodivergent child is to view that person as a whole individual; someone who is much more than the role they play as a parent or caregiver.

It's important to remember that those raising neurodivergent children are also members of their larger community where they play the role of friends, family members, colleagues, and neighbors. Each has their own preferences, needs, desires, and hopes for themselves and their family members. *Parents are people too.*

While society may place a huge amount of importance on the parental role, it's crucial to acknowledge that parents of neurodivergent children have multifaceted lives beyond caregiving. Recognizing and supporting these other dimensions can greatly contribute to making these parents or caregivers feel connected and supported by those around them. So, extend your hand. Whether it's a book club, enjoying a night out, working on a project, or simply having a relaxed get-together with friends, remember to include them in your activities.

Recognizing parent and caregiver needs

As we emphasize throughout this chapter, every human has needs. Nevertheless, many of us grow up without a complete understanding of these needs or how to fulfill them. We may mistakenly believe that we can handle everything independently or incorrectly assume that no one can offer the support we require. Those same sort of assumptions also apply to parents raising neurodivergent children. Consider this: If you were raising a neurodivergent child in a world that often doesn't fully comprehend your child and their needs, would you naturally assume that others would understand your own needs as a caregiver? Probably not.

REMEMBER

The truth is that you'll never understand the needs of a parent or caregiver without asking them first. So, asking is a great place to start.

There's a bit of a catch here: When you ask someone, "What can I do to help?" a common response from the person in need might be, "Nothing" or "I'll let you know." This may be due to uncertainty, a reluctance to burden others, feeling overwhelmed by other matters, or various other reasons.

A more effective way to provide support is to be specific and proactive in your offers. Rather than saying, "Let me know if you need anything," you can say, "I'm here to assist with grocery shopping or cooking meals," or "I can look after the kids for a few hours if you need some personal time." Offering specific help makes it easier for the person in need to accept because they don't have to figure out what they need; it's already on the table.

TIP

If you don't know where to start, that's okay! The needs of a parent or caregiver are as unique as they are, and no one can read their minds (including you). The following four areas of needs are common to parents and caregivers of neurodivergent children. Use these bullet points as a conversational starting point:

» **Emotional support:** When you notice a parent or caregiver feeling stressed or emotional while discussing their child's needs and family challenges, a wonderful way to support them is by empathetic and non-judgmental listening. While practical advice and resource connections can be valuable, humans often crave someone who can simply listen and understand. Start here.

» **Practical support:** Offer to help with specific tasks (organizing carpools, grocery shopping, babysitting, and so on). If your children attend the same school, consider helping by sharing crucial school updates, especially if the caregiver is swamped with meetings and messages. If you're their workplace manager, offer flexibility in handling work-related projects.

» **Advocacy support:** Individuals raising neurodivergent children are constantly required to advocate for their family's needs, which can be draining. You can ease their load by assisting in their advocacy. How? Advocate for a deeper understanding of neurodiversity in your shared social and workplace settings. Advocate for the accommodations that families need to integrate into community life. Actively practice inclusion in your friends group, your child's school, your workplace, and your community. This simple act of including a parent or caregiver as a regular part of your social circle goes a long way in advocating for families and their needs.

» **Empowerment support:** Go beyond addressing someone's immediate needs; offer to help them get to where they want to go. Everyone has aspirations, after all. What does a parent or caregiver want for them and their child? How do they hope to achieve it? Offer to be there for them on their journey and look for practical ways you can help out. When someone takes an interest in helping us grow, that can be hugely empowering.

WARNING

If you're a neurodivergent person supporting someone parenting a neurodivergent child, keep this in mind: Many neurodivergent individuals tend to express empathy by directly solving problems they observe. There's absolutely nothing wrong with that. In fact, it's great! Yet, neurotypical individuals often need someone to simply listen to them first for a sense of connection and understanding. Listening is an empathetic way to accommodate this need. If you do spot a solvable barrier, ask the person if they'd like assistance before jumping in (for more on this, see Chapter 11).

Helping caregivers navigate a diagnosis

Of course, one of the first points where parents and caregivers of neurodivergent children need support is when a child is diagnosed with a neurodivergent condition. You see, when a child receives a diagnosis, their families are often caught off guard. That's normal. After all, who receives a diagnosis and says, "Thanks, doc. I was expecting that. No need to tell me more!"?

What isn't normal is parents receiving compassionate, informed guidance right away. Many families are not given the information, understanding, and support that they need upon their child receiving a diagnosis of a particular neurodivergent condition. "When my child was diagnosed autistic in 2009, I was handed a brochure and DVD expounding the benefits of behavioral therapy, and told if I wanted a good outcome for my child's future, then I would need to commit to 25 to 30 hours a week of one-on-one therapy, with the ultimate goal being to make my child indistinguishable amongst his peers," writes Dr. Emma Ward for *Reframing Autism*. "Just writing those words today makes my heart hurt at the ignorance and ableism that was our introduction to autism."

The experience recounted by Dr. Ward, herself an autism researcher, is similar to that of many parents whose child is diagnosed with a neurodivergent condition. Parents and caregivers are told that their child is "broken" or "not normal" and they are seldom provided a road map on how to secure the understanding, accommodations, and support that their child and their family needs.

Instead, many families are left to wade through information on their own. This is in addition to facing a community unaware of neurodiversity, potential stigma, long wait times for services, and needing to repeatedly advocate for their child's needs. This process may also involve making significant changes to their home, lifestyle, and routines. That's a lot for a parent or caregiver to face.

ADVICE FOR HEALTH CARE PROFESSIONALS

As a health care professional, whether you're assessing a condition or delivering a diagnosis, your interactions with a neurodivergent child and their family are crucial for their well-being and growth.

- **Sharing information and feedback:** When sharing a diagnosis or delivering feedback about a neurodivergent condition to a parent or caregiver, be sensitive. After all, they may be unprepared for the conversation. Listen empathetically, provide context, then share your observations honestly. Reassure the family that

(continued)

(continued)

neurodiversity is a normal part of the human condition and focus on the things that can help their neurodivergent child thrive.

- **Recognize your own limitations:** Regardless of the quality of your medical education and training, it's unlikely that you were educated about neurodivergent conditions from a neurodiversity perspective. In the past, these conditions were often viewed through a deficit lens, but science now shows that this approach is outdated. Instead of trying to "fix" neurodivergent conditions (an endeavor no medical professional or scientist has achieved), focus your care on enabling the neurodivergent child to thrive as a neurodivergent person and on guiding the family toward accommodations and support that can help them thrive.

- **Advance your professional development:** It's natural for there to be a lag between current scientific understanding and the day-to-day practice of medical professionals. If you're a medical professional, you can bridge this gap by seeking training from reputable organizations specializing in neurodiversity. This not only enhances your professional growth but also improves the care you offer to families. Connecting with neurodivergent colleagues is another way to gain a contemporary understanding of neurodiversity. Professional organizations such as Autistic Doctors International — an organization of autistic doctors and medical students — are great places to start (check out the appendix for more).

While you may not play a role in the diagnosis of a parent or caregiver's child, you can help them navigate what comes next. In the rest of this chapter we discuss practical ways you can do just that.

Providing Support

Parents and caregivers face numerous barriers in finding the right supports and services they need. There is a mountain of conflicting information, community services, opinions, conspiracy theories, and snake oil, which every caregiver must navigate — largely all on their own. This takes a lot of time, consumes a lot of attention and energy, and can leave the caregiver feeling unsupported and disconnected from those around them. And it can lead to a lot of anxiety when having to choose the information and services that can best meet their child's needs.

REMEMBER

Parents and caregivers shouldn't have to do all of this alone. Whether you are a friend, a doctor, a service provider, or an educator, increasing your own understanding of neurodiversity and the needs of neurodivergent conditions can help you provide caregivers with the support that they need.

Providing caregivers with better information

Did you know that excess sugar causes attention-deficit/hyperactivity disorder (ADHD), or that vaccines cause autism, or that a lack of effort is the root of dyslexia? Well, none of these claims are true (excess sugar doesn't cause ADHD, vaccines don't cause autism, and dyslexia has nothing to do with a lack of effort). Nevertheless, these misconceptions have been voiced and echoed by well-intentioned individuals. Sadly, the public sphere has been inundated with so much inaccurate information about neurodivergent conditions that it leaves many parents and caregivers feeling frustrated, angry, and bewildered.

Many parents of neurodivergent children grapple with information overload from health care professionals, the media, and their communities regarding neurodivergent conditions and parenting. They encounter conflicting advice, potentially exploitative services, and struggle to locate reliable neurodiversity resources. They may not be aware of supportive research, local groups, or where to access affirming resources.

Helping parents and caregivers of neurodivergent children connect with the information, supports, and services they need is like giving them a compass and a map — it reduces anxiety, takes out the guessing, and helps them figure out the best path for them and their family.

TIP

You don't need to be an expert in neurodiversity to make a positive impact. Here are some practical ways you can connect families with the information they need:

>> **Be inquisitive:** Ask the parent or caregiver what they need and what specific resources or information would be helpful.

>> **Share knowledge:** If you come across valuable resources or information, share it with the parent or caregiver. It may be just what they've been searching for.

>> **Connect to resources:** Introduce them to local support groups, online forums, or social media communities where they can find like-minded individuals and helpful advice. (If you're looking for a place to start, check out the appendix.)

Connecting parents and caregivers to neurodivergent adults

Neurodivergent children grow up to become neurodivergent adults. Learning from their lived experiences fosters understanding and acceptance, allows parents to ask questions they may otherwise be afraid to ask, and provides families with hope, knowledge, and resources they may otherwise have never found.

In many major cities, you can find social and support groups for adults with specific neurodivergent conditions. In addition, similar online communities are easily accessible as well. These communities primarily serve as a means for neurodivergent adults to support one another, but they can also be a valuable resource for individuals seeking understanding and support for their own neurodivergent child (check out the appendix for more).

Incorporating Families into Daily Life

Whether it's intentional or not, neurodivergent families with children are often left out of their communities. Here are some ways to seamlessly involve neurodivergent families in your daily life:

>> **Promote inclusive practices:** Advocate for practices in your community, schools, places of worship, and workplaces that enable families with neurodivergent children to be included.

>> **Plan inclusive activities:** Organize inclusive gatherings with sensory-friendly options, accessible venues, and flexible schedules. Promote play among neurotypical and neurodivergent children, and educate all kids about neurodiversity, acceptance, and friendship.

>> **Collaborate in planning:** When it comes to planning school activity, sports league, or community project, invite neurodivergent families to fully participate. Collaboration helps build bonds and mutual understanding.

>> **Offer flexibility:** Be understanding and flexible when making plans or accommodations, considering the specific needs of the child.

>> **Respect boundaries:** Be mindful of boundaries and respect choices families make when it comes to accepting assistance or discussing their situation.

REMEMBER

Seek to understand the person you would like to support. Listen to them, understand their needs, and offer support. This way, we can ensure that all families, including those with neurodivergent children, can fully participate in our communities.

Chapter **24**

Shaping Society

I f you want to build a house, or anything for that matter, it helps to have a plan. Sure, you could wing it, but having a well-thought-out plan ensures that every step is clear, resources are efficiently used, and potential problems are anticipated and mitigated, leading to a smoother and more successful project. If we really want to improve the lives of neurodivergent people, and the amazing families who love them, our efforts must go beyond what we can do for ourselves. We need to build a neuroinclusive world that works for everyone. This chapter offers a blueprint on how to build that world, starting with a firm foundation.

TIP

A quick word on the voice we adopt in different parts of this chapter. In some places, we, as the author team, address everyone, neurotypical and neurodivergent alike. In other places, John and Khushboo, as actually neurodivergent individuals, address the neurodivergent readers in particular.

Laying a Firm Foundation

The foundation of building a neuroinclusive society are neurodivergent people themselves. After all, they know their lives and what they need best.

Centering neurodivergent people

Centering neurodivergent people refers to actively seeking and prioritizing the input and experiences of neurodivergent individuals in all aspects of societal development. This approach involves including neurodivergent individuals in policymaking, educational reform, workplace design, and community planning. It places neurodivergent people at the very center of creating the very policies and practices that affect them.

In schools, a neurodiversity-informed approach means developing curricula and teaching methods with input from neurodivergent students and educators. In workplaces, this approach is about creating inclusive environments and practices informed by neurodivergent employees. In communities, it means ensuring public spaces and services are accessible to all cognitive styles. Essentially, a neuroinclusive society recognizes neurodivergent individuals as experts on their experiences, leading the way in building a more inclusive society.

Constructing a common-sense approach

Centering the end user in design is not just the ethical thing to do but also the most effective approach. Including neurodivergent people in everything about them — from conception to final design — leads to policies, products, and services that are more inclusive and better suited to a diverse range of users. It's a common-sense approach, and one we should follow.

"Nothing About Us Without Us" helps us all

The motto, "Nothing About Us Without Us" is a simple yet powerful idea used by many neurodivergent people. It means that neurodivergent folks get a say in decisions about their lives, making sure their needs are met.

At first, this motto may sound a bit strong, even aggressive. Some people think it means that those using it believe they always know better than parents, health care providers, or researchers. But that's not the case. "Nothing About Us Without Us" can better be interpreted as *let us help you out!*

This motto isn't about control; it's about collaboration. It's about seeing those who want to help and saying "Hi! I see you're working toward some important goals. Let's work together. I have insights and tips that can help you reach these goals faster and more efficiently. This way, we can save time and resources, and get some things done."

Ensuring all voices are involved

We each have our own perspective, but we're not building a neuroinclusive future for just ourselves. Creating a neuroinclusive future is a group effort. Collaborating with people who have different experiences and viewpoints helps us tackle problems from multiple angles, spot barriers we may have missed, and come up with new ideas and solutions. Effective collaboration means listening to a wide range of neurodivergent voices from different cultures, languages, experiences, gender expressions, and neurotypes. We should also consider the views of parents, spouses, and families. After all, we're working to create a world that includes everyone.

Creating a Solid Frame

It's sometimes said that a well-built house has "good bones," which means the basic structure and foundation of the house are strong and sturdy and provide a solid starting point for any improvements or renovations. In the following sections, we look at what can ensure our efforts at building a neuroinclusive society are strong and sturdy.

Exercising our advocacy

Building a neuroinclusive society means we, as neurodivergent people, need to advocate for ourselves. It's like announcing, "I'm here, and my experiences are important." By speaking out, we highlight our needs and contribute to a world that appreciates diverse brain wiring. This opens doors to greater understanding and support. It's about using our voices to drive real change.

Developing our leadership

As we build a neuroinclusive world, developing our leadership skills is key. Leadership isn't just about being in charge. It has many forms, including those who like to work behind the scenes or a team. We can all become leaders. A good leader communicates clearly, listens well, inspires trust, fosters teamwork, shows empathy, and guides people toward shared goals. That's a person we all should hope to be.

REMEMBER

Clear communication isn't about how you talk or write. It's about understanding your own views and being respectful and understanding of others. Everyone communicates differently (more on this in Chapter 6), so effective communication can look different for each person.

TIP

To lead on issues that matter to you, start by discovering your passions and what you'd like to see improved. Then, consider how your unique viewpoint can contribute. Amplify your voice by joining groups, writing about your experiences, or simply talking about those things with friends.

For those wanting to help, it's key to provide leadership opportunities and supportive environments for practice. This involves mentorship, constructive feedback, and valuing their unique perspectives and strengths. Encouraging neurodivergent people to lead projects or teams, and providing necessary resources and support, aids their leadership growth.

Installing Networks

This is the point in the project where things start to get exciting. It's a wonderful feeling to realize that we don't have to go about building things on our own. Lots of folks have a stake in what we're creating.

Connecting our communities

We need to connect individuals with various neurodivergent experiences, such as autism, ADHD, and dyslexia, to build a neuroinclusive world. By uniting, we can share different perspectives and coping strategies, enhancing our collective understanding and support. Furthermore, joining forces with the wider disability community is crucial. It strengthens our advocacy for rights, accessibility, and inclusion. This collaboration not only deepens our understanding but also propels meaningful change, which benefits everyone in both the disability and neurodivergent communities.

REMEMBER

It's vital to include neurodivergent people with intellectual disabilities in our work. Historically, they've often been left out, but involving them benefits us all. We need to use all of our perspectives and experiences to make this world a better place.

Involving parents (and vice versa)

It's only natural for neurodivergent folks and those who parent neurodivergent kids to have different perspectives and experiences at times. It's sort of like the tension that many families experience when kids grow up and begin to form their own views of the world. But this doesn't stop us from loving, respecting, and listening to each other. We're still a family; that means a lot.

If we look past these small differences, we find that we all share similar goals, needs, and hopes. We all want ourselves and our families to be accepted, understood, and supported. We all strive for a world where we can be ourselves and have our needs met. And we all hope for a world that values and includes everyone. The more we see, empathize, and provide each other with support, the better off we'll all be.

TIP

If you're neurodivergent, actively include and support parents in your efforts to create a better world. And if you're a parent or caregiver of a neurodivergent person, do the same for them. And for those neurodivergent folks out their raising neurodivergent kids, help us both out along the way!

Working with allies

If we want to build a world that better meets our needs, then working with allies is key. Allies are people who may not share our experience but still believe in our goals. They're folks who lend support, spread our message wider, and use their influence to advocate for the changes we need. A really great ally is one who is willing to listen, learn, and offer support. They don't need to be experts in everything about us, nor should we expect them to be. "We don't make change with perfect allies, but with real ones," said Aubrey Blanche, an expert in designing effective systems and teams, in a 2017 article published on Medium.

Beginning to Build

We can also make the world more neuroinclusive by changing places where we spend lots of time, such as work and school. By sharing our ideas and experiences, we can make these environments better for everyone.

Modernizing schools

To make schools more neuroinclusive, we can encourage teacher training on neurodiversity and learning differences. Introducing tools such as audiobooks and visual math aids can help students who interpret letters and numbers differently. Creating quiet spaces for those sensitive to noise and involving neurodivergent students in decisions is key. We can also advocate for flexible teaching methods, such as visual aids and hands-on activities, to cater to different needs.

A great example of such modernization may be found at Chico State University, where Associate Professor Josie Blagrave and Career Advisor Betina Wildhaber have significantly improved the experience of neurodivergent students. And their approach isn't really that radical; they involve students in decisions about programs and services that impact them and work to develop their leadership skills. The result has been a campus where neurodivergent students and staff feel supported. This made Chico State a national leader on neurodiversity and elevated the school as a beacon of neuroinclusion for individuals and families in the surrounding community.

Modernizing workplaces

Workplaces should work for everyone, including people who think differently. In the following sections, we look at what we can do to shape where we work into places that support neurodiversity.

Expanding workplace networks

A key step in building neuroinclusive workplaces is creating, expanding, and nurturing our professional networks. This opens and expands communication and collaboration between us and our allies. It also helps us find connections and resources we may not have discovered otherwise. And work-related networking isn't just for those in office jobs; it's also incredibly important to those who work in trades and other fields. Builders, hair stylists, maintenance workers, and others in hands-on professions benefit greatly from networking.

For many neurodivergent individuals, the traditional customs and superficial nature of typical networking can be challenging. However, you don't have to network in that manner! Rather than seeing networking as just a business deal, aim for authentic connections. Look for people you can learn from and think about ways you can support them. Besides, neurotypical people often value the honesty and authenticity that neurodivergent individuals bring to networking. You just be you!

TIP

To build your network, start with small, practical steps. Get to know others at your workplace. You can also attend events and join online communities related to your field — platforms such as LinkedIn can be great for this!

Engaging with employee resource groups

Creating employee resource groups (ERGs) can be gamechangers. These groups offer a space where neurodivergent employees can be themselves, learn from each other, and sharpen their strengths. And they just don't help neurodivergent employees but lots of other colleagues too.

When Chris Williams cofounded the Neurodiversity Community at Square (the company now known as Block), he did so by bringing together colleagues who were neurodivergent as well as coworkers who parent neurodivergent children. As the autistic father of neurodivergent kids, his inspiration was influenced by his daughter Cassidy, whose autism diagnosis eventually led to his own. "She inspired me to learn, she inspired me to self-realize, she inspired me to seek my diagnosis, and now she's inspiring me to stand tall and make my own truths plain for others to see," he noted on the social networking site X in 2019.

Initially, the ERG founded by Williams and his colleagues started with just a few members but quickly expanded to over 1,000. Its rapid success highlights the benefit of connecting neurodivergent employees and allies in the workplace. Suddenly, both parents and neurodivergent people found themselves in a network of colleagues who understood the intricacies of neurodivergent life. That meant this diverse group of coworkers could advocate for shared needs from different angles, as well as lend each other mentorship, understanding, and support. That's a great model to follow!

Shaping workplace policy and culture

Neurodiversity ERGs also shape workplaces by including neurodivergent views in workplace policy discussions, which in turn improves workplace culture by helping companies understand and support the needs of neurodivergent employees and customers. This creates a more attractive environment for neurodivergent people, helping companies attract and keep neurodivergent talent. It turns out that's hugely important, as companies attempting to recruit neurodivergent talent without first creating a supportive environment often fall flat.

Eric Garcia, the Senior Washington Correspondent and Bureau Chief for the *Independent*, urges companies to focus internally before starting neurodivergent hiring programs. "Before you can start a neurodivergent hiring program, you need to clean your own house first," advised Garcia, who is both an autistic and an ADHDer, during a 2023 LinkedIn Live session titled "Neurodiversity For Dummies: A Community Conversation." He emphasized the importance of building a neuroinclusive workplace culture. When these hiring initiatives do launch, Garcia believes they should be guided by those with firsthand experience. "As well-intentioned as some of these neurodiversity hiring programs are, I think that they need to be led by neurodivergent people themselves."

Building effective nonprofits

Building effective nonprofits to serve neurodivergent people is essential. And to be effective, neurodivergent people must be allowed to lead and direct their efforts. That doesn't mean those of us who are neurodivergent need to go it alone. There's lots of allies to help us along the way. But when neurodivergent individuals lead on issues that are about them, they're able to apply deep insight and firsthand knowledge that drives meaningful change.

TIP

A group that says it represents a specific neurotype, but doesn't have many people with that neurotype in its leadership, is not an effective nonprofit.

Conducting Accurate Research

Shaping better research that focuses on helping neurodivergent people, rather than trying to change them, is key to building a more neuroinclusive world.

Understanding the priorities

When it comes to research priorities, neurodivergent folks and their families repeatedly ask for the same thing: areas of research that can improve their day-to-day lives. This includes focusing on areas such as assistive technology, practical accommodations for school and work, better access to health care, bolstering mental health, navigating life transitions and aging, strengthening support networks for individuals and families, overcoming employment barriers, and the like. Unfortunately, those aren't the priorities of most researchers. For instance, in the United States, only 2 percent of autism research funding each year focuses on improving the long-term lives of autistic people.

Daniel Hodges, cofounder of the Peaces of Me Foundation, which provides connections and resources to disabled people and their families, shared another common frustration — the lack of research of those with multiple life experiences. During a LinkedIn Live session titled "Neurodiversity For Dummies: A Community Conversation," Hodges pointed out the negative consequences that creates, stating, "There's a very strong possibility that my ADHD was missed for years due to my blindness. That's actually not uncommon among those of us who have physical or sensory disabilities and happen to be neurodiverse."

Understanding research barriers

Researchers often face barriers that prevent them from effectively studying neurodivergent conditions. Understanding these research barriers is essential to creating more accurate and impactful studies that truly benefit neurodivergent people. The following are common barriers researchers encounter while studying neurodivergent conditions:

>> **Funding priorities:** Donors and institutions often direct funding toward causes they find interesting or prestigious. That means those researching neurodivergent conditions who hope for financial backing often must design their research around what funders want rather than what the neurodivergent person needs.

>> **Academic pressure:** Universities often see researching the causes of neurodivergent conditions as more prestigious than studying what can help neurodivergent people. As a result, many researchers feel pressured to avoid focusing on research that could provide real help.

>> **False assumptions:** Researchers studying neurodivergent conditions often have good intent but also inaccurate assumptions about the conditions they study, resulting in research that misunderstands neurodivergent experiences.

>> **Poor study design:** False assumptions often lead to poor study design. Study design is the plan for how to conduct research, including data collection, analysis, and interpretation, and good study design is vital for ensuring research results are accurate and helpful.

Composition of research bodies

In scientific research, oversight bodies and research associations are key institutions that guide the direction of studies in a specific field. They set research priorities, ensure ethical standards, and often provide funding and support for projects, shaping how research is conducted and what areas are explored. Historically, these groups overseeing research into neurodivergent conditions have not included neurodivergent individuals in the decisions they make, which can lead to missed priorities, false assumptions, and wasted resources. But that's beginning to change.

More neurodivergent people are becoming scientists and studying their own conditions. Their personal experience not only helps their research but also positively influences other scientists as well. Research groups and oversight bodies are starting to see the value of having neurodivergent scientists in leadership roles. This is important because when neurodivergent researchers help make policy decisions within their research field, it ensures the research is relevant, respectful, and accurate. This greatly benefits the neurodivergent community they study.

Turning the Power On

Talking to elected officials about neurodiversity is crucial. Informed officials can then create better laws and policies for neurodivergent people in schools, workplaces, and communities, which may lead to improved education, more job opportunities, and a more accepting society. Some ways we can engage our elected officials include scheduling meetings with them, writing letters, sending emails, and organizing with others. And don't worry about bothering them. These officials are meant to listen to us; that's their job, and our voices can influence them to enact meaningful change.

You're more powerful than you think

Every year in the United States, lobbyists spend a staggering $4.1 billion trying to sway the decisions of elected officials. While that fact may seem disheartening at first, it actually highlights how important you are. You see, elected officials don't just make decisions on their own; they're heavily influenced by the voices around them, including lobbyists pushing agendas and constituents like you sharing your needs. And although it may seem tough to compete with campaign donations and lobbying efforts, when pressed, elected officials admit that their constituents' voices and opinions are what influence them the most. "Nothing scares a politician like a constituent," says John from our author team, who spent years working in politics.

Using your power

There's a catch to all this — you actually have to participate. "Unless people engage with their elected officials and show up at the polls, the issues they care about are less likely to be addressed," said Steve Lieberman, a former United States Senate staff member and disability advocate who is autistic, on the social networking site X in 2018.

Your voice on the local level

Engaging with your elected officials is usually easiest at the local level. They represent fewer people than national politicians, so they're more accessible. Also, local politicians often focus on *constituent services,* which means they help people like you solve specific problems in your community. For example, a school board member could actively step in to secure accommodations for a neurodivergent child if their school is uncooperative, while the office of a local council member may be able to help a neurodivergent adult find resources for independent living.

Your voice on the regional and national level

Local officials aren't the only ones interested in your needs, but higher-level government officials often have more people to serve. For instance, in the United Kingdom, a borough councilor may represent 1,500 people, while a Member of Parliament may serve up to 100,000.

Engaging regional and national representatives, sometimes joining up with other constituents helps ensure your voice is heard. That's why neurodivergent-led groups such as the Autistic Self Advocacy Network (ASAN) exist. Groups like ASAN provide neurodivergent communities with the resources and connections they need to effectively engage their representatives. So, you don't have to go it alone!

TIP

Remember to engage with public policy bodies too. In the United States, every state has a State Council on Developmental Disabilities, as required by Congress. These councils aim to shape state policies on cognitive differences. They not only welcome your input but also have members and staff ready to help you navigate and access government services.

Renovating How We Provide Therapy

We can't just imagine our way to a better life. Activities such as exercise, healthy eating, sleep, and strong social connections act as vital external supports that can help to enrich our lives. So can mindfulness, checking up on our health, pursuing our passions, and continuing to learn. And at various points in our life, we may also include all sorts of professional therapy.

Unfortunately, therapy for neurodivergent folks hasn't always been practiced with neurodivergent people in mind. We know that services that work for others don't always work for us, but those who provide us therapeutic services often don't know that. It's time to change that, both on the personal level and on a larger scale.

On the individual level

It is important to be honest with your therapeutic provider. Let them know who you are and how you think. If they don't seem to get it, you may want to look somewhere else. While providers may not know everything about your particular neurodivergent experience, they should take a neurodiversity-informed approach (see the appendix for more).

At the industry level

Modernizing therapy for neurodivergent people goes beyond individual efforts; it demands large-scale, systemic change. Here's how we can do that. First, we should train therapists and health care workers in neurodiversity principles, teaching them to understand and respect different neurological experiences.

Including neurodivergent people in creating these trainings is a crucial part of that. Next, therapy offerings should focus on boosting neurodivergent people's unique strengths, not just making them fit typical norms. Finally, we need campaigns to educate everyone about neurodiversity to reduce stigma and increase acceptance. Taking these steps can create a more inclusive and supportive therapy environment for neurodivergent individuals.

Making Sure Everyone's Work Is Seen

When covering any group, the media should include voices from that group itself. Stories about a community should feature its own members' experiences and views. This guarantees authentic, respectful coverage, avoids stereotypes, and lets the group narrate their story. It's a basic rule of journalism. Yet, coverage of neurodivergent people often misses the mark, with reporters sourcing comments about us rather than talking to neurodivergent people ourselves.

It doesn't have to be this way! Journalists strive for accuracy, so with proper guidance and awareness, the portrayal of neurodivergent people in the media can improve significantly. Whether you're a neurodivergent person or ally speaking with the press, or a journalist covering neurodivergent topics, here's a thumbnail guide on how to ensure accurate coverage.

>> **Talk to us (not about us).** When covering neurodivergent topics, speak directly to those who live it. For example, in a story about ADHD, don't just rely on quotes from neurotypical researchers, pharmaceutical companies, or family members. Talk to us.

>> **Avoid tokenizing.** For instance, when reporting on policies for dyslexic learners, make sure to include dyslexic policy experts. Don't just source non-dyslexic experts and then interview a dyslexic kid for balance.

>> **Always check your sources.** When an organization says it represents a neurodivergent group, check. For example, many groups claim to represent autistic people, but not all hire autistic staff. Some may hire a few autistic people but don't include them in leadership roles.

7

The Part of Tens

Explore ten concepts that can help you better understand neurodiversity and the role you can play in helping neurodivergent people thrive.

Discover ten ways to recognize and embrace neurodiversity to enrich the collective human experience and celebrate the diversity of minds in our world.

Chapter **25**

Ten Key Concepts about Neurodiversity

Neurodiversity is both a concept and a social movement. It is nuanced and multifaceted, and our understanding of it is evolving. The ten concepts we highlight in this chapter provide a foundation for understanding neurodiversity and the role you can play in making the world a better place for all people, regardless of their neurological profile.

Neurodiversity Is Normal

Neurodiversity is a normal and inherent part of the human experience, just as natural as any other human variation. When you meet other people, you see that they have different heights, hair textures, skin tone, voice pitch, and eye color. They have many other differences as well that vary from person to person.

There's no reason to expect that the human brain is the same across all people either. Of course, you can't see another person's brain or look inside your head to examine your own. Instead, brain variation is expressed through how we think about things and how we behave — such as how we communicate, learn, socialize, and how we process the world through our senses.

REMEMBER

Our brains are as unique as we are. However, people with similar brain traits are grouped into categories called *neurotypes*. While most belong to the same neurotype, research indicates that 20 percent or more of people fall under other neurotypes such as ADHD, autism, or dyslexia.

Neurodiversity has been a part of us as long as the human family has been alive. It turns out that society needs a mixture of neurotypes, much like nature needs biodiversity. Neurodiversity benefits our collective resilience, adaptability, and plays a key part in our biggest breakthroughs and ideas.

A Shift Is Needed in Our Perspective

Our modern understanding of neurodiversity calls for a shift in how society views neurodivergent conditions. Older ways viewed such conditions as disorders, deficits, or something to be fixed. Thankfully, this perspective is changing, and neurodivergent people are beginning to be understood for who they are as a whole.

REMEMBER

Neurodivergent people are more than the challenges they face. They have unique ways of thinking and experiencing the world that not only benefit them when supported, but which also benefit us all.

This is not to deny the challenges neurodivergent people face. However, society is beginning to recognize that many of the challenges neurodivergent people face are not due to their inherent traits, but to how society responds to them. In essence, it means that the difficulties neurodivergent individuals encounter often arise from societal barriers, such as a lack of accommodation or understanding, rather than something inherently problematic about their neurodivergent characteristics.

By addressing these societal barriers, we can reduce the challenges neurodivergent people face. An example is the expanded use of assistive technology that enables those who don't physically speak (which includes a significant number of autistic people) to communicate with those around them. This technology breaks down communication barriers, enabling neurodivergent individuals to express themselves and engage with others effectively.

REMEMBER

In shifting our perspective, we can acknowledge and respect everyone around us as normal regardless of their neurological traits, their abilities, or their support needs.

Self-Advocacy Is Essential

Whoever you are, understanding and speaking up for your needs is essential. It's what enables us to be understood, set boundaries, gain resources, and form more honest and supportive relationships. *Self-advocacy* is a term that is used to describe those moments of our life when we are able to understand and champion our needs. Asking a spouse to help you carry in groceries, working with your manager to find a flexible schedule that works for you both, and letting your friends know that you need a little alone time to recharge are all examples of self-advocacy.

In the context of neurodiversity, self-advocacy is key for thriving. It enables individuals to grow their autonomy, articulate their needs, assert their rights, and gain necessary support. Most crucially, it helps communicate that their needs are just as ordinary and important as anyone else's.

When neurodivergent people exercise self-advocacy, those around them benefit as well. This positive impact extends to families, who gain a clearer understanding of their loved one's needs; schools, where educators can better tailor their teaching methods; workplaces, which become more inclusive and accommodating; friends, who learn how to offer meaningful support; and society at large, which becomes more equitable and informed. Overall, self-advocacy enriches multiple facets of life, creating a ripple effect of benefits.

REMEMBER

Encouraging self-advocacy in neurodivergent individuals is vital for their personal growth and well-being. Parents and other key figures can play a crucial role in this by providing the tools and the safe space needed for these individuals to articulate their needs and preferences. This empowerment not only aids in their own self-discovery but also fosters a more supportive and understanding environment for everyone.

People and Families Need Better Support

Neurodivergent individuals and their families often face a lack of adequate support, whether in educational settings, health care, or the workplace. This shortfall not only hampers the personal development and well-being of neurodivergent people but also represents a missed opportunity for society. By providing targeted support and accommodations, we enable neurodivergent individuals to contribute their unique skills and perspectives. For those raising neurodivergent children, better information support can significantly reduce stress and improve the overall quality of family life.

Investing in such support structures is not just ethical; it's also practical. It leads to a more diverse and inclusive society, fosters innovation, and promotes overall social cohesion. In other words, better support for neurodivergent individuals and their families is a win-win for everyone.

Compassionate Curiosity Is Key

Compassionate curiosity is the practice of approaching someone's unique experiences and traits with both empathy and a genuine desire to understand. Instead of making judgments or assumptions, you ask open-ended questions and listen attentively, aiming to grasp what it's like to walk in a person's shoes.

For a neurodivergent individual, this approach can be empowering and validating and make them feel seen and valued. The same is true for those raising neurodivergent children. On your end, exercising compassionate curiosity not only broadens your understanding of neurodiversity but also enriches your own emotional intelligence. It fosters a deeper connection between both parties, contributing to a more inclusive and empathetic community.

REMEMBER

Compassionate curiosity cuts both ways — from neurotypical toward neurodivergent people and vice versa. Both self-advocacy and advocacy need the platform of compassionate curiosity!

Accepting People for Their Differences

Accepting someone for their differences means embracing the unique qualities that make a person who they are, even if those qualities don't conform to societal norms or expectations. If differences are seen as detractors and others feel irritated with having to deal with them and start avoiding them altogether, would the individual feel psychologically safe enough to perform well? You know the answer! Accepting differences is all about appreciating the full range of human diversity, rather than trying to fit everyone into a one-size-fits-all mold.

REMEMBER

Only when we can accept others for who they are can we give individuals the space needed to perform at their best.

Including Others for Their Abilities

Including someone for their abilities means recognizing and valuing the unique skills, talents, and experiences a person brings to the table. It's about appreciating what someone can do, rather than focusing on what they can't.

This isn't a new concept. On our author team, Ranga and Khushboo were raised in the Indian culture where ancient teachings declare, "may *all* beings in the world have happiness and peace!" Many cultures teach the idea of "treating others as you would like to be treated."

Including neurodivergent individuals and recognizing their unique abilities is another win-win. By recognizing and supporting their strengths, we empower neurodivergent people to thrive. Simultaneously, society benefits from a wider talent pool, fresh perspectives, and the understanding that each individual contributes to our collective progress.

Expanding Awareness of Neurodiversity

Expanding awareness of neurodiversity brings about numerous benefits. When people and society see neurodiversity as normal, it fosters acceptance, empathy, and a deeper understanding of diverse neurological profiles. It reduces stigma and encourages the inclusion of neurodivergent individuals in all aspects of life, from education and employment to social interactions.

This shift in perspective not only enriches the lives of neurodivergent individuals, allowing them to flourish in environments that value them, but also enhances society as a whole. It promotes diversity of thought and innovation, leading to more creative solutions and a more inclusive, compassionate, and harmonious community for everyone.

Supporting Identity and Community

Identity and community play pivotal roles in the realm of neurodiversity, significantly influencing how neurodivergent individuals perceive themselves and engage with society. Instead of perceiving themselves as "lesser versions" of what society expects, many neurodivergent individuals embrace their true selves as complete and valuable in their own right. This shift in perspective is empowering and encourages self-acceptance, allowing them to live authentically and challenge stereotypes and misconceptions.

Neurodivergent individuals often find and connect with others like them, both offline and online. Here, they form community where individuals can support one another, share experiences, and learn from each other. These communities play a crucial role in the lives of neurodivergent individuals. Within these communities, individuals often discover coping strategies, resources, and a sense of belonging that can be challenging to find elsewhere.

Supporting neurodivergent communities benefits both the individuals within them and society as a whole. For individuals, communities serve as vital support networks, boosting their mental well-being, resilience, and self-advocacy skills. They also encourage a positive sense of identity and pride in neurodivergent traits. On a broader scale, they promote an inclusive and empathetic society, reducing stigma as neurodiversity awareness increases.

Making Society Accessible for Everyone

Our modern world has been designed with certain groups of people in mind, often neglecting the diverse needs of others. However, our growing understanding of neurodiversity prompts us to reconsider and update our societal systems to accommodate not just a few but everyone.

When things are designed to work for a wide variety of neurotypes, this is called *neuroinclusive design.* Think of creating quiet and calming spaces for people who may get overwhelmed by noise, making websites that are easy to navigate for everyone, or providing information in different formats, such as video, audio, pictures, and words, so that everyone can understand it.

Neuroinclusive design benefits both neurodivergent individuals and society. For individuals, it creates a more accommodating and supportive environment, reducing stress and improving overall well-being. Society benefits from the increased participation and contributions of neurodiverse individuals, fostering innovation, diversity, and social inclusivity.

REMEMBER

Inclusive designs aren't just good for neurodivergent people; they benefit everyone.

Chapter **26**

Ten Ways to Help Neurodivergent People Thrive

iving a neurodivergent life can be a rich and wonderful experience. It often comes with unique perspectives, creative talents, and a deep connection to the world. However, it also entails facing difficulties in a world that may not always accommodate neurodivergent needs.

Many challenges stem from societal norms that often don't align with neurodivergent traits. When others fail to grasp the neurodivergent person's needs or perspectives, neurodivergent people may feel excluded and misunderstood. Recognizing and embracing neurodiversity not only benefits neurodivergent individuals but also enriches the collective human experience by celebrating the diversity of minds in our world. In this chapter, we look at ways to circumvent these barriers.

Understand Neurodiversity as Normal

The most effective way to support a neurodivergent person is to begin by embracing the idea that neurodiversity is a normal part of life. Just as two people may have different ways of solving a problem or answering a question, neurodiversity reflects the natural variation in how our brains function. If you can acknowledge the differences in communication between you and a friend or the unique work styles of you and your colleagues, then you can recognize that neurodiversity is entirely normal.

We can't observe each other's brains as we can other differences such as hair color or height. However, we express variations in our brain through how we think, communicate, socialize, and interact with the world. A neurodivergent person's way of experiencing the world may differ from yours, but that's normal. Just as we've come to understand variations in the human family, we should also understand neurological variations as ordinary.

REMEMBER

Neurodivergent people are just like everyone else. They want to be understood, accepted, and treated as normal. Some differences may not make sense to you at first, but that's okay. Take time to understand the person you are interacting with. Include them in your life, your workplace, your family, and your circle of friends.

Appreciate Everyone's Differences and Strengths

Everyone has challenges and everyone has strengths. Neurodivergent people are no different. The strengths that come from neurodiversity have been benefiting society for as long as the human family has been around. It turns out that we need people who think and experience the world differently.

Understanding the strengths and differences of various neurodivergent conditions enables you to respond compassionately and knowledgeably, empowering neurodivergent individuals to thrive in any setting. Often, beliefs about these conditions are outdated or misinformed, necessitating a reevaluation of how we perceive and interact with people who have them. By gaining insight into and valuing the unique qualities and perspectives of others, you can create an inclusive environment that benefits not only neurodivergent individuals but also everyone you engage with.

REMEMBER

Every person is different. Just because someone has a neurodivergent condition, it doesn't mean that they possess all strengths and challenges commonly associated with that neurotype.

Practice Compassionate Curiosity

To support the neurodivergent person effectively, approach them with compassionate curiosity. *Compassionate curiosity* is all about approaching someone's unique experiences and traits with empathy and a sincere desire to understand them better. Instead of jumping to conclusions or making judgments, you engage by asking open-ended questions and truly listening, aiming to gain a deeper insight into what it's like to be in their shoes.

Think about how you approach people, especially new people whose way of being seems unfamiliar to you. Do you approach them with compassionate curiosity? If you're unsure, take a look at each term separately:

>> *Compassion* is a profound awareness of and empathy for someone's feelings, whether they're going through suffering, experiencing joy, facing challenges, or achieving success. It involves engaging in conversations and relationships with kindness and empathy. The goal is to understand the other person's emotional state and experiences without downplaying or disregarding them.

>> *Curiosity* is the innate desire to learn and explore new things. When you're curious, you actively seek knowledge and understanding instead of merely accepting what's given to you. It means having a sincere interest in comprehending the other person's perspective, their life experiences, and their emotions.

Compassionate curiosity goes beyond mere curiosity for knowledge. It involves a sincere intention to understand others deeply, empathize with their experiences, recognize their emotions, and build connections founded on mutual respect and understanding. Engaging in compassionate curiosity can enhance your communication skills, minimize conflicts, strengthen relationships, encourage personal growth, and facilitate effective self-expression. This practice includes active listening, posing considerate questions, and being receptive to viewpoints that may diverge from your own. It's about fostering an environment where open, honest, and empathetic conversations can flourish.

TIP

When you make the effort to understand where a neurodivergent person is coming from it leads to better discussions, deeper understanding, and greater possibilities.

Believe in Neurodivergent People

One of the simplest yet most powerful ways to support neurodivergent individuals is to believe in them and their experiences. Many neurodivergent people feel that their genuine challenges are mistaken for excuses or attempts to avoid responsibilities. This misconception can have a profound impact on neurodivergent individuals, especially if they've shared their diagnoses and are seeking support.

When a neurodivergent person shares their difficulties, it's essential to listen with the awareness that their minds function differently in understanding the world around them. This means that what may be easy for you may be highly challenging for them, and what seems like a minor change can trigger significant anxiety. These differences may sometimes be perceived as overreactions or unreasonable requests, but open communication and proactive planning can often help find collaborative solutions. This approach not only validates and understands neurodivergent individuals but also demonstrates the practice of compassionate curiosity, considering their differences to create a more inclusive environment for everyone.

Rethink How You View Accommodations

Every human uses accommodations to get through life. If you are listening to this text as an audiobook, it's providing you an accommodation — either for your busy schedule, convenience, learning style, or pure enjoyment. Even if you're reading this book the traditional way, that too is an accommodation. You *could* spend your time and money traveling to meet our author team so that we could share with you our insight directly. But who has that time? Instead, this book serves as an accommodation for you.

All of us have a pretty good understanding of the ideal conditions that allow us to communicate, socialize, learn new things, or focus on an important task. When it comes to meeting the accommodation needs of neurodivergent individuals, it's about recognizing that these needs aren't exceptional but rather a part of creating an environment where everyone can thrive. By listening to and understanding these needs, we can foster a more inclusive and supportive space where neurodivergent individuals can shine and contribute their unique perspectives.

Advocate with Neurodivergent People

Advocating with neurodivergent people, rather than for them, is crucial to ensuring their voices are heard and their needs are met. By letting neurodivergent individuals lead the way in advocating for themselves, they become empowered to actively shape the support and accommodations they require.

This approach recognizes their expertise about their own experiences and preferences. It also promotes self-advocacy skills, which are essential for navigating a world that may not always understand their unique needs. Advocating with neurodivergent individuals fosters collaboration, mutual respect, and a more inclusive society where everyone has a say in creating environments that work for them.

Work to Include Everyone

Including neurodivergent people in all aspects of life is not just about fostering diversity; it's about harnessing the full spectrum of human potential. And it's also about getting to know more deeply some pretty amazing people.

TIP

Including neurodivergent people in friend groups, classrooms, workplaces, families, and communities is important for creating an inclusive world. Here are some simple steps to make it happen. First, take the time to understand and appreciate their unique qualities and challenges. Second, be patient and give them the space they need to express themselves. Third, listen carefully and ask questions to better understand their point of view. Fourth, be flexible and open to different ways of doing things. Finally, offer your support and encouragement when they need it. By following these steps, we can all contribute to a more inclusive and supportive environment for neurodivergent individuals.

Help Neurodivergent People Overcome Barriers

Imagine living in a world where everyone can easily get the help they need to do well, but for you, it's much harder because of how your brain works. For example, imagine needing to do a lot of complicated mental hurdles just to get a job or find a doctor. Or imagine having to fill out lots of paperwork that takes a long time to get help for your everyday needs. This is the reality for many neurodivergent people. They often feel as if there are so many obstacles in their way of securing the things they need that they'd rather hide their challenges than ask for support when they're already struggling.

To help neurodivergent individuals overcome these barriers, we can take several steps. First, we can work to make information and services more accessible and easier to understand, reducing the complexity of processes such as job applications or accessing health care. Second, we can create supportive environments where neurodivergent individuals feel safe to disclose their needs without fear of judgment. This can involve promoting open communication and actively seeking their input in decision-making. In addition, we can educate ourselves and others about neurodiversity to foster understanding and empathy. By taking these actions, we can strive to break down the unnecessary barriers that hinder neurodivergent individuals from accessing the opportunities, services, and support they deserve.

Educate Others about Neurodiversity

We unfortunately live in a world where there is still a lack of awareness and understanding about neurodiversity. Many people hold misconceptions about neurodivergent individuals and how they navigate the world. Educating others about neurodiversity is essential to promote understanding and inclusion. By sharing knowledge and dispelling misconceptions, we can create a more empathetic and accepting society where neurodivergent individuals are valued for their unique perspectives and contributions.

TIP

When educating others about neurodiversity, the best approach is to do so with compassion and empathy. Rather than calling people out or shaming them for not knowing enough, exercise the same compassionate curiosity discussed earlier in this chapter.

Practice Universal Design

Universal design means making things work well for everyone, including neurodivergent people. In schools and workplaces, we can create comfortable spaces, use adjustable materials, and listen to neurodivergent individuals to support them.

In our personal lives, we can also practice universal design to benefit neurodivergent people. When planning events or gatherings, we can make sure they are inclusive and welcoming to everyone. Using clear and structured communication in our conversations helps everyone feel heard and understood. When inviting people, it's important to be explicit and make sure they know they are welcome. By considering the needs and preferences of neurodivergent individuals in our everyday interactions, we create a more inclusive and supportive environment for all.

Appendix A

Neurodiversity Resources

None of us can build a neuroinclusive world on our own. Thankfully, there are many resources ready to provide you the information, understanding, and support that you need.

TIP

The network of knowledge, connections, and support is always growing. To see the latest resources, visit www.pivotdiversity.com/resources.

Communities

Communities are sources of valuable support, resources, empowerment, and a powerful force for advocacy and change. The following is a short list of communities for you to consider participating in.

AASCEND www.aascend.org

ADDitude www.additudemag.com

ADHD Babes (U.K.) www.adhdbabes.com

ADHD Foundation (U.K.) www.adhdfoundation.org.uk

Autistic Women and Nonbinary Network www.awnnetwork.org

Autistics 4 Autistics (Canada) https://a4aontario.com

Believe:NeuroDiversity (Australia) www.believe-nd.org

British Dyslexia Association (U.K.) www.bdadyslexia.org.uk

Children and Adults with ADHD https://chadd.org

Depression and Bipolar Support Alliance www.dbsalliance.org

Dyslexic Advantage www.dyslexicadvantage.org

Dyspraxia Foundation https://dyspraxiausa.org

International Dyslexia Association https://dyslexiaida.org

International OCD Foundation https://iocdf.org

ION Institute of Neurodiversity (Switzerland) https://ioneurodiversity.org

Kaleidoscope Society www.kaleidoscopesociety.com

Let's Talk LD www.letstalkld.org

Little Lobbyists www.littlelobbyists.org

Made By Dyslexia www.madebydyslexia.org

Mindroom (U.K.) www.mindroom.org

Minds of All Kinds www.mindsofallkinds.com

National Council on Independent Living https://ncil.org

Neurodiversity Foundation (Netherlands) www.ndfnd.org

Ourtism www.ourtism.com

Peaces of Me Foundation https://peacesofme.org

Reframing Autism (Australia) https://reframingautism.org.au

Spectrum Women (Australia) www.spectrumwomen.com

SuperTroop (U.K.) www.supertroop.org

Twainbow www.twainbow.org

Employment

Many organizations' mission is to support individuals seeking employment and to advocate for changes in hiring and management practices toward greater neuroinclusion in the workplace. In addition to the resources listed here, be sure to investigate local government agencies as well for available employment-related services.

Autistic At Work https://autisticatwork.com

Employer Assistance and Resource Network https://askearn.org

Integrate Advisors www.integrateadvisors.org

Job Accommodation Network https://askjan.org

Neurodiversity Academy (Australia) www.neurodiversityacademy.com

Neurodiversity Hub (Australia) www.neurodiversityhub.org

Neurodiversity in Business (U.K.) www.neurodiversityinbusiness.org

Neurodiversity in the Workplace https://nitw.org

Neurodiversity Media (Australia) www.neurodiversitymedia.com

Neurodiversity Pathways https://ndpathways.org

Neuroinclusion Hub (Australia) www.neuroinclusionhub.com

Neurotalent Works www.neurotalentworks.org

Pivot Neurodiversity www.pivotdiversity.com

Informational Resources

The following is a set of resources with highly useful information for parents, medical professionals, academics and researchers studying neurodivergent conditions, and individuals seeking neurodiversity-informed therapy.

For parents

Ausome Training (Ireland) https://ausometraining.com

Autistic Parents (U.K.) www.autisticparentsuk.org

I Can Network (Australia) https://icannetwork.online

Inclusive World https://inclusiveworld.org

Learn From Autistics https://learnfromautistics.com

NeuroDiverCity (Singapore) www.neurodivercitysg.com

Neurodiversity Education (Netherlands) www.neurodiversityeducation academy.org

Neurodiversity Is Normal www.neurodiversityisnormal.org

Neurodiversity Training International (Ireland) https://neurodiversity-training.com

Ollibean https://ollibean.com

Thinking Person's Guide to Autism https://thinkingautismguide.com

Understood www.understood.org

For medical professionals

Autistic Doctors International https://autisticdoctorsinternational.com

Autistic Physical Therapist https://iriswarchall.com

For academics and scientific researchers

AASPIRE https://aaspire.org

A.J. Drexel Autism Institute https://drexel.edu/autisminstitute

Autistic Research Committee (INSAR) www.autism-insar.org/page/insarARC

Autistic Researcher Review Board https://airpnetwork.ucla.edu/autistic-researcher-review-board

Autistica (U.K.) https://autistica.org.uk

CRAE (U.K.) https://crae.ioe.ac.uk

Neurodiversity Foundation (Netherlands) www.ndfnd.org

Participatory Autism https://participatoryautismresearch.wordpress.com

Salvesen Mindroom Research (U.K.) https://salvesen-research.ed.ac.uk

Stanford Neurodiversity Project https://med.stanford.edu/neurodiversit

The Frist Center for Autism and Innovation www.vanderbilt.edu/autismandinnovation

For those seeking therapy

Inclusive Therapists www.inclusivetherapists.com

Neurodivergent Practitioners https://neurodivergentpractitioners.org

Neurodivergent Therapists (U.K.) https://neurodivergenttherapists.com

Neurodivergent Therapists https://ndtherapists.com

Therapy Den www.therapyden.com

Therapy for Black Girls https://therapyforblackgirls.com

Practical Resources

The following resources may be helpful for supporting differences in communication, sensory processing. and accessibility.

Apps and tools

Boardmaker https://goboardmaker.com

Calm www.calm.com

Dyslexia Quest www.nessy.com

Fidget Widgets https://relish-life.com

Focus@Will www.focusatwill.com

Insight Timer https://insighttimer.com

Proloquo2Go www.assistiveware.com

Time Timer www.timetimer.com

TouchChat HD https://touchchatapp.com

Voice Dream Reader www.voicedream.com

Communication resources

AAC Institute https://aacinstitute.org

AssistiveWare https://assistiveware.com

Communication First https://communicationfirst.org

iTaalk Autism www.itaalk.org

LAMP https://lampwflapp.com

PrAACtical AAC https://praacticalaac.org

Speak For Yourself https://speakforyourself.org

Spell to Communicate https://i-asc.org/s2c-spelling-to-communicate

Tobii Dynavox https://us.tobiidynavox.com

TouchChat https://touchchatapp.com

Design resources

Hack the IEP `www.theadvocazine.com`

Inclusive Design Toolkit by Microsoft `https://inclusive.microsoft.design`

Jen White-Johnson `https://jenwhitejohnson.com`

The A11y Project `www.a11yproject.com`

Sensory resources

Chewigem `https://chewigem.com`

Fancy Fidget `www.fancyfidget.com`

Sensory Toolhouse `https://sensorytoolhouse.com`

Stimm `https://stimm.jewelry`

Index

H

habit loop, 160–161
habit replacement, 161
habits of effectiveness
  constructive response versus toxic positivity, 164
  contributing to shared success, 166–167
  focusing on things you can control, 163–164
  listening first before speaking, 166
  overview, 160–161
  prioritizing essentials, 165
  seeking mutual benefits, 165–166
  starting with destination in sight, 164–165
  valuing yourself, 161–163
Hallowell, Edward, 84
handwriting difficulties, 111, 115
harassment, 269
Harris, Johnny, 102, 105
health care, 42, 228, 270, 319–320
heart, nurturing, 162–163
Heidel, Jamie A., 67
help, asking for, 143
herd awareness, 237
Heslabeck, Russ, 106
high school, 187, 285–286
high-functioning autism, 57
Hilton, Paris, 94
Hingsburger, Dave, 128–129
hinting, 68
hiring practices, 303–304
Hodges, Daniel, 330
honoring self-descriptions, 58
Hopkins, Anthony, 30
human brain, variations in, 10–12
human condition, variations in, 9–10

human family, 8–9
Hutton, Sue, 95
hyperactivity, 88
hyperfocus, 63–64, 86–87, 145
hypomania, 123

I

ICD (International Classification of Diseases), 50, 51
icons, used in book, 3–4
ID (intellectual disability), 80, 127–129, 132
IEP (individualized education plan), 267–268
impulse control, 89, 92–93, 124, 143
including for abilities, 239–240, 341
inclusion. See also universal design
  helping to thrive, 347
  hiring practices, 303–304
  lens of, 18
  recruiting practices, 302–303
independent living, planning for, 272–273
indirect versus direct language, 66–68
individualized education plan (IEP), 267–268
information
  informational resources, 351–352
  informational support, 182
  lack of, as barrier to educators, 278–279
  providing parents with, 321
information-seeking, 257
innovation, 145, 291–294
institutionalization, 40, 41
integrated classrooms, 266–267
integrity, 146
intellectual disability (ID), 80, 127–129, 132

intelligence, 104, 116
interests, 140, 142, 146
International Classification of Diseases (ICD), 50, 51
interviews, 204, 303
intuition, 143
iTaalk Autism Foundation, 255–256

J

jealousy, 159
job roles, 39, 200
jobs. See employment
Joffe, Margaux, 86
Johnson, Steven, 291–292
joy, neurodivergent, 262–263
judgment, lens of, 18

K

Kai Syng Tan, 34
Kamprad, Ingvar, 35
Katz, Neal, 56
Kelly, Orion, 137
Kelsey, The, 312
Kwok, Rosa, 35

L

language. See also communication
  direct versus indirect, 67
  in dyspraxia, 114
  literal interpretation of, 67, 140
  speaking versus nonspeaking people, 55–56
laughter, 163
laziness, myth about, 98
leadership, developing, 325–326
learned helplessness, 160
learned optimism, 160
legal barriers, 42

Niou, Yuh-Line, 29–30
nondiscrimination policies, 297
nonlinear thinking, 59
nonprofits, building effective, 329–330
nonspeaking autistic individuals, 55–56, 82
nonverbal communication, 244
nonverbal learning disorder (NLD), 79–80
normal distribution, 233
normalcy of neurodiversity, 176–178, 232–233, 337–338, 344
normality, myth of, 37–39
"Nothing About Us Without Us" motto, 324
numbers, difficulty with, 108–110, 140
nutrition, 161

O

obsessive-compulsive disorder (OCD)
  compulsions, 120–121
  conflicts with daily life, 121
  empathy, showing, 121–122
  finding support, 122
  misconceptions about, 132
  obsessions, 120
  overview, 119
  prevalence of, 32
  strengths, 122
ongoing education, 287
online research, 138
oppositional defiant disorder (ODD), 80
Orfalea, Paul, 101
organizational skills
  in ADHD, 88–89, 91, 94, 97
  evaluating, 144
  support from other people, 242

organizing thoughts, 111
outcomes, focusing on, 164–165

P

pain tolerance, 145
parents and caregivers
  advocacy
    for academic progress and success, 268–270
    in classrooms, 265–268
    legal rights, understanding, 265
    for medical needs, 270
    needs of your family, understanding, 264–265
    overview, 264–265
    for support services, 265
  assessment after children's diagnosis, 33
  common barriers, 252–256
  coping strategies
    balancing needs of all children, 261–262
    examples of, 257
    overview, 257
    self-care, 258–259
    understanding needs of children, 259–261
    understanding your needs, 258
  disability community, support from, 27–28
  financial support, lack of, 255–256
  helping school-aged children, 189
  informational resources for, 351
  limited availability of time and energy, 256
  myth that autism is caused by, 82
  neurodivergent individuals as, 220–221

neuroinclusive society, building, 326–327
overview, 251
planning for future
  for advanced education, 271
  for employment, 271
  for financial security, 273–275
  for independent living, 272–273
  for lifelong care, 273
  overview, 270
relationships with, 214–217
support for
  better information, providing, 321
  connecting with neurodivergent adults, 321–322
  empathy, practicing, 316–320
  incorporating families into daily life, 322
  navigating diagnosis, 319–320
  needs, recognizing, 317–318
  overview, 315, 339–340
  understanding parents as people, 316–317
support network, building, 263–264
partners, relationships with, 217–219
passive communication, 206–207
passive-aggressive communication, 206–207
patience, 152, 242, 243, 245, 246
pattern recognition, 145
patterns of thinking, in autism, 58–60
Pauley, Jane, 123
perfectionism, 298–299
performance assessments, 39, 210
Perry, Katy, 121

About the Authors

John Marble is the founder of Pivot Neurodiversity, an organization empowering companies to modernize recruitment, hiring, and culture to better support neurodivergent employees. He is also a classroom instructor and training partner with Neurodiversity Pathways. A writer and speaker on innovation, workplace culture, and neurodiversity, he has served as an advisor to various policy makers and as an aide to two U.S. presidents. In 2009, he was appointed by President Obama to serve in the United States Office of Personnel Management, where he became the first openly autistic presidential staff member in American history.

Multiply neurodivergent, John credits his ADHD, autism, and dyscalculia as gifting him with a curiosity and passion to understand the world around him. This has led him to build a career centered on understanding and valuing the diverse perspectives and experiences of others. His ability to see various viewpoints proved particularly beneficial in the often-divided world of politics, allowing him to develop solutions that were inclusive and effective for everyone. John brings this same approach to his work as a teacher, consultant, public speaker, and visual artist. He is committed to promoting neurodiversity in a way that fosters mutual understanding among everyone involved, ensuring that neurodivergent people, parents, and allies are heard, empowered, and understood.

Khushboo Chabria is a neurodiversity specialist and a transformational leader on a mission to advocate for and help improve access to high-quality support services for neurodivergent individuals. Khushboo is a program manager, career coach, and speaker with the Neurodiversity Pathways program of Goodwill of Silicon Valley focused on educating and supporting neurodivergent individuals to help launch their careers and supporting organizations to integrate neurodivergent employees into the workplace through belonging and empowerment. Khushboo also sits on the board of the Peaces of Me Foundation and is involved in consulting, writing, and professionally speaking on the topics of neurodiversity, DEIB, leadership, psychological safety, mental health, and coaching. With varied experiences in supporting neurodivergent individuals of all ages and their family members, working as a therapist and clinician, studying organizational leadership, and discovering her own ADHD, Khushboo brings an interesting mix of skills and experiences to this field of work. Khushboo aims to make a meaningful impact in the world through education, empowerment, authentic engagement, and unbridled compassion.

Ranga Jayaraman is the director of the Neurodiversity Pathways program of Goodwill of Silicon Valley in San Jose, California. He is a neurotypical father of a neurodivergent son. Ranga's current passion is to empower neurodivergent individuals with the desire and ability to contribute; to find meaningful, rewarding, and sustained employment; and to help organizations in their journey to embrace neurodiversity inclusion in their workplaces. Previously, Ranga was an accomplished senior executive of digital transformation, contributing innovative solutions and services to enable and support business strategy and growth in information technology and higher education organizations. His real-life experience, executive leadership, and business background helps shape the classroom curriculum and learning experience Ranga provides to neurodivergent individuals and organizations. Ranga is driven by the core beliefs that everyone is a unique creation of the universe, fully equipped to fulfill the purpose for which they are here and that his life's purpose is to *Love All, Serve All, Help Ever, and Hurt Never.*

Dedication

This book is dedicated to all the wonderful neurodivergent people in our lives — family members, friends, colleagues, and students. We have learned so much from them. We are better humans and our lives are immeasurably richer because of them. Thank you!

John Marble: To the future generations of neurodivergent people who will think this book outdated because they understand much more than what we do now; and to my mom.

Khushboo Chabria: To my guides, my mentors, my loved ones and my community for empowering me and helping me become who I am today.

Ranga Jayaraman: To my constant guide and guru, Bhagavan Sri Sathya Sai Baba.

Authors' Acknowledgments

We want to acknowledge and thank four groups of people:

Our friends and families, for standing by us, cheering us on, and giving us the dedicated time for focusing on this important work.

Goodwill of Silicon Valley leadership, Trish Dorsey and Mike Fox, for supporting our work in the Neurodiversity Pathways program and affording us the time to work on this book. Many of the principles we have covered in this book evolved through our work in Neurodiversity Pathways.

Our students in the Neurodiversity Pathways program, for opening our heads and hearts to many insights and perspectives into the beauty and power of neurodivergent minds. We have learned more from you than we may have taught you!

Our team at Wiley: Our senior acquisitions editor Tracy Boggier for encouraging and convincing us to undertake the writing of this book; our Dummies coach Vicky Adang for helping shape the TOC and orienting us to the Dummies style; our development editor Katharine Dvorak for keeping us focused on every deadline and cheering us on when we felt overwhelmed; and our technical editor Sara Sanders Gardner for ensuring our content and tone is empowering, inclusive, and aligned with the modern understanding of neurodiversity.

John would like to additionally thank Ryan Alvarez, Cody Arnold, Pedro Belo, Aubrey Blanche, Cindy Cramer Blanchard, Karen Chin, Derek Gerson, Todd Elmer, Jon Heilbron, Thomas George, Alex Kotran, Rohan Mahadevan, Cass Nelson, Dave Noble, Rachel Payne, Mark Perriello, Toby Quaranta, Dick Sincerbeaux, Noah Zoschke, and the entire Tzaba family for their support. He also would like to express his appreciation for the examples provided to him by Josie Blagrave, Jen Emira, Chris Ereneta, Jen White-Johnson, Knox and Kevin Johnson, the Brooke and Lars Olsen family, Gabor Pap, Amitesh Parikh, Anne Pinkowski, Scott Robertson, Shannon Des Roches Rosa, Leo Rosa, Eve Wanetick, Joni Whitworth, and Betina Wildhaber. He also acknowledges the foundational research and writing of James Baldwin, Lydia X. Z. Brown, Barb Cook, Eric Garcia, Sue Fletcher-Watson, Edward Hollowell, Sara Luterman, Dylan Matthews, Damian Milton, Haley Moss, Morénike Giwa Onaiwu, Yenn Purkis, Steve Silberman, Alice Wong, and others. He is also deeply appreciative of the advocacy and leadership of AASCEND, the Autistic Self Advocacy Network, Julia Bascom, Sascha Bittner, Rebecca Cokley, Lois Curtis, Andrew Eddy, Finn Gardner, Elizabeth Grigsby, Frank Kameny, Judy Heumann, Margaux Joffe, Steve Lieberman, Sheraden Nicholau, Ross Pollard, Ed Roberts, Kayla Smith, Chris Williams, Elaine Wilson, Bob Witeck, and Stella Young; as well as thanking Ari Ne'eman for telling him to write something practical as the world didn't need to read another autistic memoir.

Khushboo would like to express heartfelt gratitude for the unwavering patience, understanding, and support extended by her parents, Vijaya and Anup Chabria, her brother, Manish Chabria, and her loyal companion, Emily, during the writing of this book. Special thanks are also extended to friends and supporters: Patricia Clariza, Dr. Thelmisha Vincent, Warfred Cabanes, Akanksha Aurora, Tseten Dolkar, Mahathee Chetlapalli, Dr. Jeffrey Lee, Daniel Hodges, Brian Hilliard, Linda Fisk, and the leadership teams at Wiley, Goodwill of Silicon Valley, Peaces of Me Foundation, and the Ed.D. Organizational Leadership Program at UMass Global for their unwavering encouragement and support. Lastly, sincere appreciation is extended to every neurodivergent person whom Khushboo has had the pleasure to meet, connect with, support, and empower in this lifetime. It is an honor to serve this community. Of course, this journey could not have been undertaken without the invaluable contributions of Khushboo's neuroinclusive author team who have not only empowered her as a neurodivergent individual but have also significantly shaped and expanded her understanding of neurodiversity in both profound and subtle ways.

Ranga would like to express his deep gratitude to his wife, Shamala Jayaraman, for all of her unwavering faith in him and unflinching support that have been his bedrock and backbone through many ups and downs through the last 40 years; loving appreciation to his children, Aparna and Prashant, for all the joy and love they have given him and for all the learning opportunities they have provided him; and special thanks to Jose Velasco and Thorkil Sonne for opening his eyes to autism and all that can and needs to be done to make life better for neurodivergent people like his son. Lastly, Ranga's deep appreciation to his coauthors, Khushboo and John, whose deep collaboration made this book a reality.

Publisher's Acknowledgments

Senior Acquisitions Editor: Tracy Boggier

Managing Editor: Ajith Kumar

Project and Development Editor: Katharine Dvorak

Technical Editor: Sara Sanders Gardner

Production Editor: Pradesh Kumar

Cover Images: © Vectorig/Getty Images; Iryna Spodarenko/Getty Images